One Last River Run

Square-timber cribs in Ottawa, 1898. In the background can be seen the footings for the Alexandria Bridge, which was then under construction. Notice the cookhouses and the large tug ready to pull the rafts to Québec City.

ONE LAST RIVER RUN

RON CORBETT

GENERAL STORE PUBLISHING HOUSE
499 O'Brien Road, Box 415
Renfrew, Ontario, Canada K7V 4A6
Telephone (613) 432-7691 or 1-800-465-6072
www.gsph.com

ISBN 978-1-897508-27-5

Design and layout: Magdalene Carson / New Leaf Publication Design
Printed by Custom Printers of Renfrew Ltd., Renfrew, Ontario
Printed and bound in Canada

Library and Archives Canada Cataloguing in Publication

Corbett, Ron, 1959-
    One last river run / Ron Corbett.

ISBN 978-1-897508-27-5

    1. Logging--Ottawa River Valley (Québec and Ont.)--History.
2. Ottawa River Valley (Québec and Ont.)--History. I. Title.
SD540.3.C2C67 2008   386'.5097138   C2008-906224-8

Front cover photo: Tom Stephenson is silhouetted against rain clouds while steering the timber crib along the Ottawa River before landing just upstream from Gillies Old Mill in Braeside on Monday, June 23, 2008. (Darren Calabrese / Courtesy of the *Ottawa Sun*)

Back cover photos: Tom Stephenson and Algonquin College student Trevor Slack work on the raft in the pole yard. (Dana Shaw)

Second Printing

Table of Contents

Prologue   7

PART ONE

1   Pembroke, Ontario, April 2008   11

2   Ottawa, Ontario, April 2008   16

3   Pembroke, Ontario, April 2008   20

4   Deep River, Ontario, May 2008   27

5   May 2008   33

6   Deep River, Ontario, May 2008   37

7   Early June 2008   41

8   Mid June 2008   46

9   Pembroke, Ontario, June 2008   51

PART TWO

10   Castleford, Ontario, June 23, 2008   63

11   June 23, 2008   67

12   Quyon, Québec, June 24, 2008   71

13   June 25, 2008   79

14   Dunrobin, Ontario, June 27, 2008   85

15   June 27, 2008   90

16   June 28–30, 2008   98

Epilogue   107

About the Author   111

For Julie,
who also believes in crazy ideas

Prologue

By the condition of the box, it had been some time since anyone had bothered to look inside. There was dust on the lid. Then a soft, air-sucking sound when it was lifted off, like some old seal being broken.

As I peered inside, a librarian explained the rules. I could shoot using a digital camera, but no flash. There was a photocopy service available—the cost would be forty cents a page. The box and its contents had to stay in the room. No pens. It's recommended you use white gloves.

On the research floor of the National Library, there was no other sound but the librarian talking. People sat at small cubicles, wearing white gloves, peering inside their own boxes. Those who had arrived early had cubicles facing north, with a view of the Ottawa River. The late-arrivals were clustered by the main doors.

"Will you be making any more requests today?"

I turned from the window to see the librarian standing behind me, positioned near a box of white gloves I could not hope to miss.

"No," I answered. "It will take most of the day just to see what's here, I should think. Thank you."

She nodded and left. Before taking away the box, I stared one more time out the window. I could see the north channel of the Ottawa River sweeping its way around Victoria Island. The Chaudière Falls—or what's left of them—would be about 100 metres to the west of us, just out of view. They would have set out from below the falls. I could walk there at lunch, if I wanted. That's how close we were.

I closed my eyes and imagined how it must have looked:

Tiberius, all of eighteen, standing by the banks of the Ottawa River, his pants wet and stiff from walking through the early morning dew. The river in front of him would have been loud and frothy. On the other side were the soaring cliffs that blocked the southern horizon.

Wrightville, July 1806

It had taken the boy some time to become accustomed to this new world. The winters were nothing like Massachusetts; the wind, the bitter cold—so cold on some days it made the very limbs of the trees moan and speak—nothing could have prepared him for it.

He had been here six years. Had come with his father, along with thirty-six other men, five women, and twenty children. Although their lives had been hard and the work repetitive, no one had complained. Tiberius was not surprised. Everyone here wanted the same thing:

Land.

It was all that his father and his friends had talked about when they used to gather on the sweeping lawn of First Church on Sunday afternoons. We need more land. We will never be as successful as our fathers, if we toil our life away on forty acres.

So they had come to a new world where the land around them was so great and enveloping, it was like living with a shadow person. Someone who never left your side. Who was always there. That's the way the land around here felt. Omnipresent.

But there had been problems in this new world. It was why he was here so early in the morning, while in a perfect world he would have been getting ready for a day's work in the fields.

They needed money. That was the plain truth of it. The settlement's first crop of wheat, brought to the harbour of Montréal the previous summer, had been a disaster. The selling price had not covered the expenses of the trip.

Some in the settlement had wanted to bring another crop of wheat to Montréal, but his father had argued for a different plan. Better to fail anew, he had said, than make the same mistake twice.

Eventually he persuaded the men, and so Tiberius and every other man in the settlement spent the latter part of the winter and all of the spring felling trees and squaring them. Seven hundred and eighty-eight trees they brought to the banks of la Grande Rivière du Nord, beneath the great falls the Indians called Asticou.

In the distance, he saw his father and three other men approaching. He started to walk towards them.

I opened my eyes, grabbed a pair of white gloves and walked to a cubicle. On the side of my box was a label that said Wright Papers. Below that were the words: Raft Books.

A square timber raft on the Ottawa River.

PART ONE

Loggers squaring a white pine.

CHAPTER ONE

Pembroke, Ontario, April 2008

Dana Shaw sat in the office of Herb Shaw and Sons, struggling to keep his mind on the paperwork in front of him. It was not an easy task this morning. The weather had something to do with that—a warm spring morning, after a winter that missed sliding into the record books by less than fifteen centimetres. One last medium-sized storm and they would have made it.

Outside his window, the highway between Pembroke and Petawawa was muddy and bordered with black ice. Transport trucks rumbled by in low gear, stopping for the red lights in front of the Ontario Provincial Police station and the turn-off to the Trans-Canada Highway. There was a hitchhiker the other side of the lights. It was a sloppy spring day, and people were on the move.

The weather had something to do with his lack of concentration. Shaw would have loved to have been out there with the transport trucks, making his way to the Trans-Canada; maybe he'd take a run up to Deep River that afternoon and inspect some land the company was thinking of buying.

As he returned his attention to the purchase orders in front of him, his cousin Johnny walked by the open door of his office. He was back early from the sawmill and Dana suspected he was restless, too.

John was the president of Herb Shaw and Sons; Dana, four years his junior, was vice-president.

Shaw's was the oldest family-owned lumber company in Canada, founded in 1847 by the cousins' great-great-grandfather John Shaw who came from Inverness, Scotland to start logging on Lake Doré, near Eganville. He built a water-powered sawmill and gristmill (the stones and wheels brought directly from Scotland) and before long, he was one of the most prominent lumbermen in the Upper Ottawa Valley. While other family logging operations were bought out or slowly faded away with time, the Shaws kept at it, expanding their timber limits and finding a niche market in hydro poles.

God love hydro poles. Every few years there would be a flurry of news stories extolling the wisdom of burying hydro lines instead of stringing them overhead like clotheslines, but nothing, thankfully, ever came from them. The cost was prohibitive. The infrastructure already existed. So utility companies kept buying hydro poles, and Herb Shaw and Sons—which had some of the finest red pine in Canada—kept adding years to its record of being the oldest family-owned lumber company in Canada.

Dana continued flipping through the purchase orders on his desk, listening to his cousin pace the floor. Yes, John was restless. He put down the purchase orders, stood up and made his way to his cousin's office. He stood in the open door and said:

"It's a crazy idea."

"One of the craziest I ever heard," agreed John.

"I don't see how it can be done."

"It can't be done," said John. "Where would you even begin?"

Dana nodded and walked away, moving in the direction of a coffee pot in the far corner of the accounting room. So it was agreed. A crazy idea.

Still, it was hard not to think about it. You get one phone call with someone pitching you the most off-the-wall idea you've heard in years, and you remember it. Talk about it a little. Maybe there was a full moon or something. But you get two phone calls the same week from people who don't know each other pitching the same idea, and what are you supposed to make of it?

Dana took his coffee and walked back to his desk. Coincidence? Sure, it could be coincidence, but in Dana's experience coincidence was running into your wife in the same grocery aisle or getting money the same day a boat you had your eye on went on sale — that was coincidence. This, well, this was just bizarre.

He and Johnny had never even considered it, and they would be the ones to think about it, he guessed, which probably explained why people had contacted them. They had done crazy things in the past. Like the last unofficial log run down the Madawaska River — they had done that — two weeks on the river, running logs down a waterway that hadn't been open in years even though it was still the best

way to get the wood out. In return for letting them do that, they had helped the municipality build a wharf in Horton Township. It was a wonder they didn't drown that summer.

So maybe it made sense, getting these phone calls. Although two in the same week — one from the City of Pembroke, one from the Canadian Forestry Association — that was a bit freaky. Dana grabbed his coffee and walked back to his cousin's office, standing again in the open doorway.

"We certainly have the wood."

"We have the wood," agreed John.

"It would just be a matter of building it."

"Which, I think, has been done once in the past 100 years. They might as well be asking us to build a catapult."

"We have pictures," said Dana. "Maybe we could figure it out that way."

"I couldn't do it," said John. "Unless you can think of someone who can build it, it's just a crazy idea that isn't going anywhere."

"You're right, it's crazy."

He walked back to his office and for the rest of the day paced the floor, looking outside and thinking of spring, walking into John's office from time to time, agreeing with him that building a square-timber crib was just about the craziest idea the cousins had ever heard.

Tom Stephenson walked quietly around his kitchen, trying not to wake his wife while he made breakfast. He had lived in this house for thirty years and could do the job blindfolded — second cupboard to the right, find the instant oats. Cutlery in the upper drawer. Bowls and juice cups above the sink. He made no attempt to turn on a light.

Hundreds of square-timber cribs, tied together and ready for the trip to Québec.

He had always loved the early morning. Even when he was a child, getting up before the sun rose so he could get ready for the walk to the schoolhouse in Perth, he had loved that time of day. Loved the way the first rays of sun came into the kitchen, the wooden table slowly getting brighter as he ate his porridge, as if the sun itself was burnishing the wood.

Later, when he started working in the lumber camps, he felt the same way. While other men complained about having to leave their beds on a January morning, Stephenson jumped up to greet the day. He would be the first in the cookhouse, then the first outside, enjoying the few minutes when he would be alone in the bush, and the rest of the men were finishing their meal, the sun starting to appear over a tree line to the east, the ice and snow creaking and whispering around him.

Old habits, he guessed. You reach a certain age, and things you've been doing your whole life, you don't question them much anymore. Gather up enough years, and it would be like questioning the colour of your eyes, or the straightness of your hair. What would be the sense?

He made his porridge, poured a juice and walked outside. It had been a long winter; for sheer snow, the worst since 1971, which was a winter he remembered well. There was so much snow in the bush that year, not even the horses could get into some of the camps. You were completely cut off. And this in the days before computers and SAT phones, when "completely cut" off actually meant something. That year he thought the snow would never stop falling.

Go through it once, and you won't be surprised the next time. That's the way experience is supposed to work, and while Stephenson knew plenty of men who continued to be surprised by their lives—shocked when their wives threw them out, surprised when they woke up with hangovers—he had never been one of them. He usually knew what to expect.

He sat on his back stoop and looked around. There was still snow in the far north corner of the yard, and he could tell, just by the way the hoarfrost looked on a pile of mud next to his garage, that the ground was frozen. It was going to be a late spring.

That didn't bother him as much as it once did. It might delay his vegetable garden and some of the other planting he did every spring in the backyard, but he wasn't a farmer. And it had been several years since he worked in a lumber camp that sent its logs to market by way of the river when a late spring would have meant something.

He ate his porridge and started planning his day. He would go to Fellowes High School that afternoon to visit their woodworking shop and see how the students were progressing on a pointer boat they were building. He had helped draw the plans. Before that he would drive to a sugar bush he co-owned near Lanark. It had been a lousy spring for the sap, so little of it running through the trees, it was a miracle they bottled anything at all. He had seen that before, too.

He finished his porridge and juice, the sun just starting to appear over his back shed, then went back to the kitchen. There he noticed, for the first time, a note his wife must have written the night before when he had been out late at the sugar camp, left for him by the toaster. He lifted the piece of paper and read it:

"Phone Johnny Shaw."

I'm speaking on the phone with Dave Lemkay, general manager of the Canadian Forestry Association:

"You need to travel the old highway," he says. "It's about five kilometres outside town, on the way to Petawawa. You can't miss it. It's right across the highway from the OPP station."

I listen to the instructions. It has been a long winter in Ottawa, and a bunker mentality has set in over the city. People shuffle through the remnants of snowbanks that look like the relics of some long-past war. No one seems happy. Everyone wants to be someplace else.

"You're going to love these guys," Lemkay continues. "Johnny Shaw Jr. is fifth generation. Same for Dana. So you're sure you know how to get there?"

"I'd pass Riverside Park. Right?"

"That's right."

"I know how to get there."

"Great. We'll see you Thursday morning then."

I hang up and stare outside. The snowbank in front of my house, which once stood three feet above the roofline of my car, has shrunk to a stub of ice running next to a sad-looking interlock-brick pathway. A squat, orange vehicle with a City of Ottawa logo on its side plows my sidewalk even though there is no longer snow.

A quick road trip down the Trans-Canada Highway. Get out of the city for a day. I would have been interested in the story if the phone call had been about nothing more that the opening of a chip stand.

But Dave Lemkay was promising something better. The start to a story I had been working on for several weeks, and which had begun to look impossible.

I stood from my desk and grabbed a file folder. Inside were photos printed the night before. On top of the stack was a picture of the Duke and Duchess of Windsor sitting atop a wooden raft. The Duchess wore Victorian finery; the Duke was resplendent in tuxedo and ribbons. There were other people on the raft, as well, and it was hurtling down a timber slide next to the Chaudière Falls. The date on the photo was 1906.

The next photo was of roughly 100 rafts tied together as a full moon hung over Lake of Two Mountains where the Ottawa River empties into the St. Lawrence River. There was a campfire burning on the looped-together rafts, a line of tents, men smoking as they prepared to bunk down for the night. In the background I could see someone playing a fiddle. It looked like a miniature city.

By the time I had finished emptying the file, I had eleven photos spread in front of me of a square-timber raft somewhere on the Ottawa River.

Build one of these today? People had not seen one on the river in 100 years. Was it even possible?

CHAPTER TWO

Ottawa, Ontario, April 2008

For several weeks before receiving the phone call from Dave Lemkay, I had been reading about rafts. As an Ottawa boy, I felt a twinge of shame the more I read, never realizing before how important these ungainly vessels were to the history of our region and the country.

We all know the canoe story, how the little vessel you could throw on your back opened up the Canadian interior, how it begat Samuel de Champlain, the voyageurs and the fur trade. How it is still the vehicle of choice in great swaths of the country, and what's more, many of us own one or have paddled one. Canadians have a giddy love affair with the canoe.

Rafts, in comparison, are poor cousins, even though these wonky wooden boats were just as instrumental in opening up Canada and just as important to our history.

We certainly can't claim ownership of a raft the way we can with a canoe. Rafts are the world's oldest boats, archeologists having found remnants of the vessels going back to the stone ages, rickety wooden structures latched together with animal skins and sharpened bones. After these stone age vessels came the mighty rafts and barges of the Nile. The floating bamboo cities of Thailand. In Arthurian England, there were probably more rafts than there were horses.

The literal definition of a raft, according to the *Oxford English Dictionary*, is "a flat, buoyant structure of timber or other materials fastened together, used as a boat or floating platform."

Accurate, but without any of the nuances of the vessel, without any of the romance or history. A raft, in many ways, was all about independence. It was cheap to build. It kept you mobile. It was the standard-bred horse, the Model T car, the vehicle-for-the-masses of another time. It was no coincidence that American writer Mark Twain, when he was looking for characters and a locale for his vision of the harmonious society, chose a young boy, an escaped slave and a raft.

The Adventures of Huckleberry Finn is populated with people living on rafts, working on rafts (the mighty cargo barges of the Mississippi) and escaping on rafts to some unexplored, exotic place downriver. It was much the same in Canada in the nineteenth century (although you tended, up here, to get off them as soon as November rolled around.)

There was, though, a crucial difference between rafts in Canada and those along the Mississippi River. Virtually all of our rafts were built by loggers who sent them downriver so they could sell the wood. These were known as timber rafts, or timber cribs, as they were often called in the Ottawa Valley.

Timber rafts have been around almost as long as rafts, themselves. Raftsmen societies were formed in Germany in the tenth century, most of its members piloting timber cribs. By the eleventh and twelfth centuries in Europe, timber rafts were a common sight on most major rivers. The Bavarian city of Munich, just one of many examples, was founded in 1185, built stick-by-stick with oak and pine brought down the Isar River by way of raft.

The people who made their living on the rafts were hard-working adventurers who passed along the secrets of the river only to family. They became secretive clans, and in the taverns of Bavaria and Prussia that were set aside for raftsmen, you would have been taking chances with your health if you tried to enter without knowing anyone.

In time, the timber from the Black Forest became an export commodity, and there was no better client anywhere in Europe than the British Royal Navy. The appetite of the British Navy for oak and pine was near insatiable. By 1801, 8,000 timber rafts were going down the Isar River every year, with most of the wood destined for England.

Then in 1806, Napoleon defeated Prussia and the French navy cut off English access to the Ems, Weser and Elbe rivers. In short order, as Napolean's troops marched their way through Eastern Europe, the mouths of the Isar, Vistula and Oder were closed.

The ensuing naval blockade of the Balkans would continue, or so it seemed for the raftsmen living there, for nearly a generation. The timber industry in Eastern Europe was decimated. And the British Royal Navy was forced to look elsewhere for its wood.

Funny how history, for all its swagger, its grand-sweeping arc, its tales of nations and kings, doesn't really function all that differently from the messy lives of mere mortals. You lay your plans. You think it through. You are careful about things. Then something comes right out of left field, something you never saw coming, and everything changes.

That's what happened in the Balkans. And the British Navy's misfortune changed the destiny of a small outpost halfway across the world.

Wrightville, July 10, 1806

Philemon saw his son walking by the banks of the river and stopped to wait for him. Tiberius had already been below the falls to inspect the rafts. They would start to load the oak staves on the rafts that afternoon, assuming Tiberius had found everything ready for the journey.

As he waited for his son to approach, Philemon looked around. He never got tired of looking at this land, at the western horizon that might as well be the edge of the known world; at the forests of pine and spruce that surely rivalled the Black Forest in Germany; at the river that roared by their small settlement, one of the wildest waterways he had ever seen.

The difference between this land and the land he left, it was night and day. In Woburn, Massachusetts, there were nearly 200 years of settlement history, slow-moving rivers, large stone homes and a winter you rarely dreaded. Many would have preferred that, but hundreds of acres of wilderness where you could create something anew, that was more to his liking.

He was forty-six and had always been an adventur-ous soul, enlisting in the Revolutionary War when he was fifteen, fighting at Bunker Hill, among other battles. After the war, he married Abigail Wyman, and as was the custom at the time, was given a forty-acre portion of his father's estate on which to raise a family. Almost immediately he felt constrained.

He started investigating land grants in British North America, and in 1797 petitioned for a grant in Hull Township, in a part of Upper Canada where there were no other settlers. The following year he made a journey to inspect the land along la Grande Rivière du Nord. He surveyed the Rideau River, the Gatineau Hills, and marvelled at the forests he walked through and the dark, rich earth beneath his feet. Surely this was the place to build an Eden-like community of like-minded, farmer-dreamers who no longer wanted to toil on their fathers' forty-acre gifts.

He returned to Woburn and convinced five other families that there was no future for them in New England. He had seen land where there was no limit to what you could build, what you could grow, what you could create.

His enthusiasm—and his financing of the endeavour, he supposed—was hard to resist. So on February 2, 1800, Wright, his wife, his seven children and the other families from Woburn—sixty-three people in total, including workers he had hired for the season—set off for their new home, eventually travelling in covered wagons on a rough road that ran beside the St. Lawrence River. The seigneury estates could be seen from the road, heading back in rectangular strips of land from the shores of the river. Then they reached Lake of Two Mountains, and the road ended. Here, they turned north where there were no estates. No road. The covered wagons travelled down the frozen Ottawa River, every passing mile taking them further from towns and villages, from the places where people actually lived.

On the first day on the river, an Algonquin hunter and his wife stumbled upon the settlers and at first the man thought he was seeing ghosts, so astonished was he to see covered wagons this far in the forest. But then he left his wife beneath a tree, turned around and escorted the party the rest of the way, taking the lead and tapping the ice with a wooden stick to make sure it would support the weight of the wagons— tap, tap, tap, moving slowly, silently, the snow billowing around the covered wagons, tap, tap, tap, all the way to the Chaudière Falls where, to the hunter's astonishment, Wright told him they had reached their destination.

It was a harrowing trip. But once they reached the end of their journey, the other families agreed with him. This was beautiful country, a forest like none they had seen before, where the air was alive with water, the sounds of waterfalls and rivers heard day and night; where the trees stood as tall as cathedral spires and the earth was dark and rich.

The best laid plans, Philemon supposed. It took six years and an inestimable amount of labour—they had built a water-powered gristmill and sawmill, cleared more than 500 acres of land, built many homes—before he realized the land held a cruel surprise. They were on a small band of fertile land running beside the river. Go two miles back into the hills and cliffs that bordered the river, and the earth was shallow and rocky. Their first cash crop, brought to market in Montréal the year before, had actually lost money.

As the leader of the struggling settlement, he was desperate for some form of commerce, some way of bringing money into the settlement. For no matter how self-sustaining the community could become, there was still no avoiding the need for hard cash to buy seed and other supplies in Montréal.

Then a trader had come up the Ottawa River the summer before and mentioned it was a shame these trees were not growing around Québec, for lumbermen down there were doing a brisk business with English merchants. The pine. The oak. They were worth money.

He thought about it over the winter until he was past the point of distraction. He needed money. His trees were worth money. How could he connect the two? Then, although it seemed absurd when the idea first came to him, a possible solution presented itself. What if they were to sail the wood to Québec?

He watched Tiberius approach, until his son stood before him

"The rafts?" he asked.

"They seem ready, Father. All nineteen of them."

He nodded. All right, it was time to tempt the gods.

CHAPTER THREE

Pembroke, Ontario, April 2008

"That's Lake Doré. Probably the turn of the (last) century."

Dana Shaw stands behind me in a large office filled with wooden furniture and brass-appointed filing cabinets. He is pointing with his finger at a framed black-and-white photo hung on the wall above the head of a receptionist.

"That was where the company started. It's on the Snake River."

I look at a picture of an Algonquin Highland lake, recognizable by its dense bush, white pine, silver birch, granite outcrops. A solitary sawmill, looking as out of place as an Inca temple in the middle of the city, is in one corner of the photo.

The instructions to get here were perfect. I arrived at Herb Shaw and Sons—five kilometres outside Pembroke, on the way to Petawawa, directly across the street from the OPP station—at a little past ten in the morning.

Waiting for me were Dana and Johnny Shaw; joining them were Dave Lemkay and Fred Blackstein from the Ottawa River Designating Committee. We grab coffees and start talking about rafts.

The last timber raft to make its way down the Ottawa River made the trip in 1908 and no commercial raft had gone down the river since. Dave Lemkay built one twenty-five years before, for the 100th anniversary of the Ottawa Valley being declared a conservation forest area, but he never sent it downriver, and he wasn't the one who actually built it. Old lumbermen did that. He would try to find them. Yvon Soucie was one of them. He might be able to help. He tended a farm though, and it was a busy time of year.

"Do you think it can be done, Dave?" asked Dana Shaw.

"Sure," said Lemkay. "You just need time, money and wood."

So for the next two hours we debate time, money and wood. This was a major undertaking being considered. These rafts were huge. Did anyone even know if it would be sea-worthy when it was built? Should you even bother trying to sail it? What would be the point, though, of building a raft if you weren't going to put it in the water?

On this point there was a vigorous debate. Building a raft was one thing. Trying to send it down the Ottawa River—where towns, dams, and government red tape had spawned like zebra mussels—that was another. Who would you even phone first to start getting the necessary approvals?

Then there was the cost. Each hydro pole was about $1,000. The raft might take thirty or more. This was a sizeable investment, just on the raw materials.

Anyway, the conversation went back and forth like that for several hours, although for most of the time I stayed silent and kept my eyes on the Shaws, deducing these were the people who were going to make this idea happen—or not. I was encouraged by their enthusiasm. Dana left the meeting once to grab a black-and-white photo hanging in some inner office: "We'll build it like this," he said, pointing to several rafts floating below Parliament Hill, circa the late-nineteenth century—"We'll just use the photo. That's all we need."

As for John, he leaned forward in his chair the entire meeting, like a man waiting to get on his feet and do something. By noon, I was starting to think the chances of building a timber raft one more time in the Ottawa Valley were better than I previously thought.

Which meant we had gone from nearly impossible to long shot. The cost. The red tape. The shortage of lumberjacks older than a century. The obstacles were daunting.

"Do you really think you can do this?" I asked Dana, as the meeting was coming to an end. To which he replied:

"I don't know. But I think we're about to find out."

Tom Stephenson hung up the phone and walked outside. Judy was turning sod in a flowerbed.

"So what did Johnny want?" she shouted to her husband.

"He wants me to help him build a raft."

"A raft?"

"That's what he says. One of the old timber cribs. He wants to build it down in the pole yard."

"Why would he want to do a thing like that?"

"I have no idea."

And he didn't. As he grabbed a rake and made his way towards his wife, it occurred to him he had never thought to ask. Johnny had just started talking, as if building a square timber raft—something Tom had only seen in photos—would be a perfectly normal thing to be discussing on a Tuesday morning. Hey, Tom. Doing anything this spring? Want to build a raft?

He started raking leaves off a flower bed, gathering them in a small pile by the fence. There was snow next to the fence, and he broke it up with his boot. The sugar camp was just about done for the season, so he had some time. He told Johnny he would drive down that afternoon and talk to him. Told him, at the same time, that the idea sounded crazy, but Johnny just laughed and said he already knew that. What time can you be here?

He guessed the Shaws would have the wood for such a thing. If anyone would have it, it would be the cousins. They dealt mostly in softwood, red and white pine, and that's what the old rafts were made from. Sometimes, they carried oak staves on board, but it was mostly pine. He had read somewhere that the old rafts could be as long as 100 feet, which gave you some idea of how big the white pine used to be around here.

Most of the first-growth white pine was long gone when Tom started logging, but every once in a while, he'd stumble upon a stand. He'd crest a hill, or round a bend in a river, and it was like walking into a cathedral—the sheer height of the trees took his breath away. Three men with their arms linked couldn't circle the circumference of one tree. The sun couldn't reach the ground. The wind moving the needles in the tallest branches couldn't be heard.

Trees like that, you'd see old photos of them, and it was a wonder they ever came down. Which maybe explained why so many men would pose in front of them once they fell. Like they were posing with fish they had just caught or something. Tough-looking men, standing with the catch of the day, and Tom would have hated to tangle with any one of them.

The most skilled of them would square the trees after they fell, and the most skilled of the skilled would then build the rafts. They did a lot of the work out on the ice of a highland lake, or on one of the rivers that ran into the Ottawa River, the rafts simply melting through one spring day, and that would be the beginning of the long journey to Québec City.

Tom kept raking leaves, breaking up the last of the snow with his boot, thinking that most of the rafts would have been out on the river by now. Pembroke would have been a zoo, so many rafts travelling down the Ottawa River, you probably could have walked across the river from the town wharf. There would have been an unholy din of shouting and singing, the taverns by the waterfront overflowing, and the wood moving down river day and night. God, it would have been quite the sight. He checked his watch. He had told Johnny he would be there shortly after lunch.

Johnny Shaw drove his truck through the pole yard, turning left and right through the stacked tiers of red and white pine. There were 42,000 poles in the yard today and more trucks coming in from Deep River that afternoon. In spring, the yard was always full, always lots of work to do, but for most of the week Johnny had been thinking about little more than wooden rafts.

He had been reading about rafts. Found some books in the library. One[1] was by Charlotte Whitton the former mayor of Ottawa. She was a Renfrew girl, her father a forestry services worker who once worked for Gillies Brothers, a lumber company headquartered in Braeside, that used to compete with the Shaws. For more than 100 years they competed, just to get the story straight. Still couldn't believe, some days, that the Gillies company was gone.

He had fallen into the habit of reading the books late at night when his wife was in bed, and the work was done for the day. He would look at photos of rafts—as many as 100 of them tied up sometimes for the trip—and the scale, the sheer size of the vessels, it seemed preposterous to him. He couldn't imagine it.

He read of how Philemon Wright opened up the timber industry in the Ottawa Valley with a crazy, boldly-going-where-no-one-has-gone-before trip down the Ottawa River that everyone at the time said was a suicide mission. No timber slides. No navigation charts. Just rapids after stretch of rapids, all the way to Québec City. He took nineteen rafts with him. And his eighteen-year-old son.

It was a daring act, and Johnny marvelled at it. Marvelled again when he read that within fifty years of that trip, square timber from the Ottawa

---

1   Whitton, Charlotte. 1943. *A Hundred Years A-Fellin': The timber saga of Ottawa 1842–1942*. Ottawa: Runge Press for Gillies Bros.

John Shaw, left, and Dana Shaw negotiate their way across a pile of white pine logs during the initial stages of making a square timber crib. (Darren Calabrese / Courtesty of the Ottawa Sun)

Valley had become the biggest single export in British North America. That crazy old New Englander had practically saved the country from an eternity of trapping beaver.

Johnny drove through the tiers of wood until he reached the northwest corner of the pole yard where they kept the uncut wood, the trees right out of the bush. He parked the truck. There was still snow in the forest that ringed the edge of the yard, and Johnny was reminded again of how long the winter had been. Seemed like it would never end. If you had been in a lumber camp during a winter like that, you probably would have talked about it forever.

He climbed out of the truck, zippering his coat as he closed the door. It had been a slow spring. Not much warmth to the days yet. Good for bringing the wood out, he guessed. Not much else.

He made his way to the trees that had come out of Alice Township the day before, white pine, about fifty of them. The bark on the pine was gnarly and covered with frost. He stood atop a tree and began walking, like a lumberjack out on the river, counting as he went.

He stopped after thirty and looked back. He figured that was what they would need to build the raft. In his research, he had discovered that the rafts were all built so they could travel down a timber slide. The length didn't matter all that much, but the width could not be more then twenty-eight feet.

How you tied all that together, he still wasn't sure. The rafts were built without nails, screws or twine. He stared out at the trees—spreading all the way to the edge of the forest—and wondered how they did it.

He jumped off a few minutes later and looked at his watch. Tom would be here shortly. He looked back at the tree he had been standing on, suddenly curious to know how much it weighed. Johnny swept his gaze around the pole yard until he saw a loader working in a far corner, then raised his arm and motioned the driver over. Let's see what we're actually getting into.

A half-hour later, when the scale in the back of the yard tipped 2,400 pounds—more than a ton—Johnny let out a slow whistle. This was nuts.

A lot of people thought the story was nuts. Right from the start.

My involvement had started the previous March, at a lunch meeting with the editor of the *Ottawa Sun*. I had being doing research, looking for ideas for a possible summer series for the paper, when I stumbled upon a small mention on the Internet about the last square-timber raft to go down the Ottawa River.

It had gone down the river in 1908, and for reasons I couldn't figure out at the time, there had never been another one. Just like that—poof—a glorious part of the city's history had disappeared.

If someone thought of building a raft and taking it down river, it would make a wonderful anniversary series, and Mike Therien agreed. I left lunch with a firm commitment from the paper for a summer series on the first timber raft to go down the Ottawa River in 100 years.

There would be just a few small problems to overcome. For starters, I needed to find someone willing to build a raft. Preferably someone with a lot of wood.

In the days that followed, I came to regret the successful story pitch. No one, it quickly became apparent, had any plans to build a raft in honour of the 100th anniversary. Most people wondered if it would even be possible. Who would have that much wood? Who would build it? A 100-year-old lumberjack?

Then if you built it, how in the world would you take it down the Ottawa River? You would need Transport Canada approval. Maybe Fisheries and Oceans. Maybe the National Capital Commission. And you want to do this when? By June?

A curator at the Museum of Civilization laughed at me. The NCC laughed at me. Amateur historians, who had written books on the logging industry in the Ottawa Valley laughed. The more I phoned around Ottawa, the louder the laughter became. Two weeks later, I was beginning to dread phoning Therien and

A square timber raft about to set off down a series of rapids on the Ottawa River.

telling him something along the lines of "Listen, we may have jumped the gun here a little."

Then I phoned the Canadian Forestry Association and spoke to Dave Lemkay. He phoned the Shaws. And suddenly I'm on my way to a raft meeting, looking for the OPP station between Pembroke and Petawawa.

But even after that first meeting—maybe it was the laughter still ringing in my ears—I doubted if we could pull this off. I had begun to appreciate the scope of the undertaking. The cost. The problems with government red tape. Long before the Victoria Day weekend, people were starting to say this might be a better idea next year—even if we did miss the anniversary.

That uncertainty continued for weeks. This was a project that always seemed on the verge of collapse. Too crazy. Too rushed. No time to phone Transport Canada. No time to worry about getting around dams, or moving a thirty-ton raft downriver, or who could pilot such a thing, and where in the world were we going!

Then a handful of adventurers just dove in. And along the way these modern-day pioneers—from Tom Stephenson to the last full-time teamster still working the bush in Eastern Ontario; the Shaw cousins to a baker-turned-miller in Deep River; a startled PR person at the Canadian Museum of Civilization to a car-wash owner in Ottawa—this odd collection of kindred souls made it happen.

As our plans took shape, I did a lot of reading myself. About Philemon Wright, for the most part, this curious man who first settled the Ottawa Valley, who named one of his children Christopher Columbus and brought another on a voyage that opened up this country, as surely as any voyage by Samuel de Champlain.

He made a raft trip down the Ottawa River and then the St. Lawrence River through the Long Sault Rapids, the Lachine Rapids, the North Channel of Isle de Montréal that no one at the time thought was possible. It was a desperate act by a desperate man, and I tried to imagine what it would have been like for him, making that journey, wondering if he had made a mistake by moving north, wondering every evening as he slept on his raft, or as he wrote in his journal, about his future and his fate.

His story made a powerful backdrop for our own endeavour. Put us to shame, to be honest, on days when we thought the task ahead was impossible. Or, ultimately, too absurd.

CHAPTER FOUR

Deep River, Ontario, May 2008

I'm bouncing down a gravel road with Dana Shaw, on our way to get one last tree for the raft. The Shaws want to use horses to fell one of the trees, and we are on our way to meet Willy O'Brien.

Willy O'Brien lives on 250 acres south of Eganville, next door to his brother Paddy and not far from his son who is also called Willy. The O'Briens have been logging the Upper Ottawa Valley since the early nineteenth century, and while much about lumbering has changed over the years (no more timber rafts, for one), the changes haven't affected the O'Brien's all that much.

The latest generations—Willy, Willy and Paddy—still work in the bush felling trees; still work as a family; still farm when they're not logging. And they still use horses to haul out the trees, "skid it out," as the O'Brien's say, even though there is no skidder.

Willy Sr. owns the horses, half-a-dozen Belgians. He learned to run a team of horses from his father, who learned it from his father and so on down the line. Willy says he may be the last, full-time "teamster" in Ontario. A few do it part-time, or as sport, but getting up every day and going out to work with the horses—he may be the last one.

"I still use horses because horses are still the best way to work in the bush," says Willy, after we

are introduced. He is sitting in the back of a pickup truck eating a homemade pepperette. It is early morning, but the O'Brien's have already been waiting for us for nearly an hour.

"Skidders are easier," he says. "That's all."

I think about that a minute—the best not being the easiest—as I look around the forest we are standing in. We are south of Deep River, not far from the Trans-Canada Highway, and it is not particularly dense. Hard maple and ash, with some nice white pine sprinkled about. It is the white pine we are after.

Behind me, a pair of Belgians snort and paw the ground. Willie Jr. gets the harness from the back of the truck. Paddy stands and stretches. We are about to start.

Philemon Wright likely would have used oxen to skid out his trees. He brought eight of the animals north with him when he emigrated from Woburn, Massachusetts in 1800.

After the oxen would have come, the workhorses, then the mechanical skidders. The skidders came along around the same time as the first chainsaws, neither work tool having a carburetor in the early years, and the sound in the forest was an unholy din because of it.

Willy O'Brien "skidding" out a white pine near Deep River. (Darren Calabrese/ Courtesy of the Ottawa Sun*)*

It's still too loud, says Willy, as I walk beside him. He hated working in the lumber camps when he was a young man. The noise of the cookhouse, the trucks coming and going all day and night. Even when you left the camps and went into the bush, there would be the skidders and the saws and the men who could never stop talking to save their souls. Like they were afraid of silence or something.

He prefers working with his family and his horses. You have more control over your life that way. It's quieter, too.

We walk through the forest, talking quietly, looking around at the trees, until O'Brien says "Johnny, what do you think of this one?"

Coming up behind us are Johnny and Dana, along with Paddy and Willy Jr. We all stare at the tree. A brief debate ensues. The tree has been struck by lightning, and some feel this should rule it out—"It's not a full tree," complains Johnny. Others, however, feel this makes the tree the perfect length for a raft—"It's less work," says Dana. Eventually the good-length-for-a-raft people win the day.

Willy Jr. starts the chainsaw. He makes a forward cut—for the direction of the fall—then goes to the back of the pine and starts to saw. He has to use a splint, before he is finished, when the tree refuses to tumble, but within ten minutes it's on the ground.

The two Belgians are quickly hooked up. Willy Sr. takes a long rein and stands behind them, off to one side, where he gives a quick flip of his wrist, a cluck of his tongue, and the horses start to walk.

It takes a while. The ground is muddy, and Willy doesn't work the horses hard. Other teamsters would have been quicker, but Willy is different. It would have been faster with a skidder too, so that's not really the point.

"Where are you taking the sticks, Dana?" he says when the horses have finally dragged the tree to the road.

"We're taking them to Deep River to get them squared," answers Shaw. "The fellow has already started. You should come up and have a look."

When Wright brought his first patch of trees to the banks of the Ottawa River, he squared them as well. This was always the second step in building a timber raft.

The wood was squared—the round edges of the tree cut away so a square centre remained—because this maximized the number of trees you could fit in the hold of a merchant ship bound for England.

Some credit Wright with designing the logger's square-ax—a tool specifically made for squaring large trees—although this seems unlikely. His historical list of firsts will have to suffice with first permanent settler along the Ottawa River and first person to build a timber raft in Upper Canada.

Willy O'Brien behind his team of Belgians. O'Brien may be the last full-time teamster in Eastern Ontario. (Darren Calabrese/ Courtesy of the Ottawa Sun.

How the notion of building a raft came to Wright, no one knows with certainty. He came to the New World to be a farmer, not a logger. He would have seen timber rafts in New England, although not many. Perhaps he rode on one during the Revolutionary War, as rafts were a common way to move troops. Still, he had certainly never attempted to build one before 1806.

He had a few problems to contend with. The Ottawa River and the St. Lawrence River were much larger, faster-moving rivers than anything in New England. The raft had to be sturdy.

Ideally, it also should be able to carry oak which was too heavy to float, but which commanded top dollar from the British Navy. He needed to build large pine rafts to carry the oak. Small wouldn't do.

Strangely, for a man as meticulous as Wright, he never left a plan for the rafts he designed. And after that, everyone learned by watching. It would be nearly 150 years before any sort of plan would be published, and then it would appear in the book by Charlotte Whitton. By then, it had been nearly half a century since anyone had actually seen one.

July 15, 1806

Philemon and Tiberius had been on the river for five days, and already they had experienced so many obstacles, it was tempting to turn around and go home. Take another stab at getting a cash crop to Montréal.

The day before, one of the cribs had run aground. They had spent most of the day trying to get the wood off the muddy shoals, and for their efforts it did not look like the raft had risen, or moved, so much as an inch. Today had been a repeat performance of yesterday's futility.

Philemon had not been discouraged. If they needed to, they would take the raft apart, stick by stick, and send it down river that way. They could put it back together someplace where the river slowed, some bay downriver. If the raft had not lifted from the shoal by midday tomorrow, that's what they would do. Anything was possible, he told his son, until it had been proven impossible.

That night they ate in front of a fire while the stars overhead shone as brightly as fireworks, one frozen moment of bright, white fireworks. There were more stars that night than Tiberius had ever seen before, or perhaps he was simply noticing them for the first time. The land around here made you sit up and pay attention, so maybe that was it.

Before they were done, the shoals might seem like the least of their problems. The Lachine Rapids on the St. Lawrence River still lay ahead—that legendary stretch of white water that kept European explorers on the other side of it for more than a century; a perilous stretch of rapids that was once the edge of the known world.

It felt strange sitting in the forest, under these stars, knowing that he was heading to a place like Lachine. He had read somewhere that the name was French for China, because people once believed on the other side of the rapids lay the Oriental kingdom. Just upriver and the other side of a sweet-water sea. Champlain believed it.

Strange, but you still didn't know what you would find if you headed far enough west. Not with certainty. It was like some rootless, make-believe world he was living in, with stars that looked like fireworks displays, with rivers that deafened your ears and rafts that were larger than a crop field. It disoriented you sometimes. Made you light-headed.

Tiberius looked back at the sky and began to think the stars—their vividness, their originality—might be a good omen for the trip ahead. He was a young man and given to such reflections. He was also his father's son.

His father, after all, had left Woburn to move his family into howling wilderness. They had renounced their citizenship, taking an oath of loyalty to a foreign king, then tap-tapped their way up a frozen river to a crazy place of waterfalls and towering trees, where the nearest neighbour was 100 miles away. His father was as rootless as the land. The year before they left New England, his youngest brother had been born. His father named the baby Christopher Columbus.

No, the Wrights did not mind a good adventure.

The O'Briens didn't mind a good adventure either. Before leaving, Willy, Paddy and Willy peppered the Shaws with questions about the raft:

Who was going to square the timber again?

Greg Merrill, he built that big log house outside Deep River. Used to be a baker in Almonte. You've met him.

Where are you going to build the raft?

In the back of the pole yard. We'll be out of the way there. We'll truck the sticks back next week, when Greg is finished.

And who is actually going to build the darn thing? I've never seen one.

We're getting Tom Stephenson. You know him. Used to work the camps on the Québec side. He lives in Pembroke.

Stephenson. Yeah, I think I've heard of him. He has a maple stand, right?

A big one. Makes syrup. And rope. Have you ever seen his rope? He has some crazy old machine he uses.

No, never seen his rope. Just heard about him.

Shortly after that, the O'Briens loaded the horses into the back of the trailer. It had been an easy day's work. Most of the wood needed to build the raft was already sitting in the pole yard. They took plastic bags from the truck and pulled out sandwiches wrapped in oiled paper. Began eating.

For several minutes no one talked. I walked around the forest, noticing how the late-spring sun was coming through the trees so it cast filigreed shadows on the ground. The shadows moved when the wind blew. It made the earth seem to move. In the distance, I could hear a flock of geese returning, and somewhere closer, cars moving down the Trans-Canada Highway.

I was looking forward to getting up to Deep River to meet Greg Merrill who would be squaring the trees. He had some wonky, back-yard sawmill, apparently, and some helpers you don't find every day. Garlic farmers for the most part. Those people are squirrely, I was told. Was looking forward to meeting Tom Stephenson as well, a man I had been hearing a lot about from the Shaws. If even half the stories were true, the guy was a second cousin to Big Joe Muffraw.

When I returned to the vehicles, the O'Briens were rolling up their oiled paper into balls and stuffing them in their plastic bags.

"It was nice meeting you," said Willy Sr. as he climbed out of his cab to shake my hand. "I wish you the best of luck with the raft."

"Do you think we're going to need it?" I asked, standing there shaking his hand, waiting for Willy's certain denial, a dopey grin spreading across my face.

"Don't rightly know," he said, before extracting his hand. "Talk to me in a couple of weeks."

A "brag load" at a lumber camp in the Ottawa Valley. Loggers often spent Sundays, their one day off, building massive loads of wood and then taking joke photos. The two horses, mercifully, were only hitched up for the photographer.

CHAPTER FIVE

May 2008

The Victoria Day weekend came and everyone involved in the raft project dispersed for the holiday, promising to meet the following Tuesday at Greg Merrill's backyard sawmill in Deep River.

Johnny Shaw spent the weekend working in his backyard, breaking up the frozen ground in preparation for planting, putting together a barbecue, cleaning up the garage. He was a man who liked moving, even on the days that were given to him for rest. He started working in the lumberyard when he was fifteen, and some days it seemed like he hadn't stopped.

Still, that weekend, in the evening, he took some time to read his raft books. Pored through the pages. Looked again at the photos he liked best — the rafts tied up below Parliament Hill; a solitary raft making its way down a set of rapids, two men hanging onto a sweep that looked to be doing little more than dangling in the water, as useless against the rushing water as a twig.

In the Charlotte Whitton book, he had actually found a diagram of a square-timber raft. It was the closest thing to a blueprint or building plan he had discovered. In the text accompanying the diagram, the loquacious former mayor of Ottawa even explained how the raft should be put together, with

traverses and pins. There was the answer right there. How to put one of these things together without any nails, screws, glue or twine.

Dana saw the diagram that weekend and was just as excited. It showed the base of the raft. The traverses on top. The sweeps. It was something to start with. He drew a copy of the diagram and brought it home, where he studied it over the kitchen table.

Monday evening, he also took out his digital camera to check the batteries and the number of memory cards he had. He had always regretted not bringing a camera when he and Johnny had embarked on the last timber run down the Madawaska River. Sure, there were no shortages of stories to tell after that trip, but he would have traded a dozen of them for one photo of Johnny falling into the river, or the logs early in the morning gathered on some bay with the mist burning off, and the sun just starting to throw beads of light across the water. Why he didn't bring a camera, he'll never understand. He wasn't going to make that mistake twice.

Across town, Tom Stephenson also spent the weekend doing chores, preparing for the coming season, all the while his mind somewhere else. The trees had been felled. The wood had been trucked

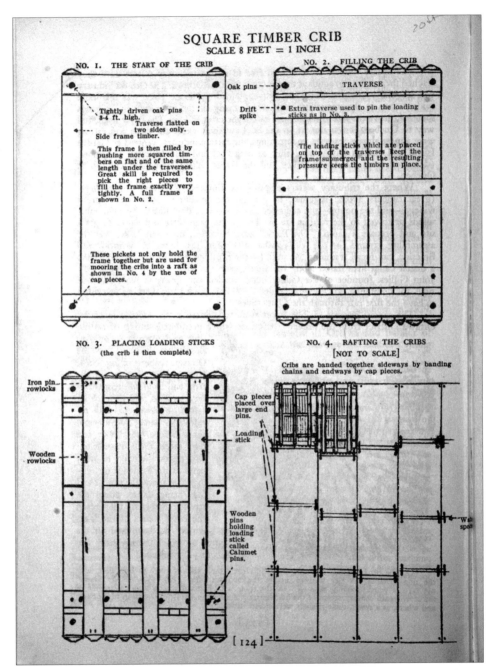

SQUARE TIMBER CRIB
SCALE 8 FEET = 1 INCH

NO. 1.   THE START OF THE CRIB

Oak pins

Tightly driven oak pins 3-4 ft. high.

Traverse flatted on two sides only.

Side frame timber.

This frame is then filled by pushing more squared timbers on flat and of the same length under the traverses. Great skill is required to pick the right pieces to fill the frame exactly very tightly. A full frame is shown in No. 2.

These pickets not only hold the frame together but are used for mooring the cribs into a raft as shown in No. 4 by the use of cap pieces.

NO. 2.   FILLING THE CRIB

TRAVERSE

Drift spike

Extra traverse used to pin the loading sticks as in No. 3.

The loading sticks which are placed on top of the traverses keep the frame submerged and the resulting pressure keeps the timbers in place.

NO. 3.   PLACING LOADING STICKS
(the crib is then complete)

Iron pin rowlocks

Wooden rowlocks

Wooden pins holding loading stick called Calumet pins.

NO. 4.   RAFTING THE CRIBS
[NOT TO SCALE]

Cribs are banded together sideways by banding chains and endways by cap pieces.

Cap pieces placed over large end pins.

Loading stick

Web spar

[124]

Diagrams of a square-timber crib, from Charlotte Whitton's book A Hundred Years A-Fellin' to the sawmill in Deep River. It was time to start sawing wood and building a raft.

Monday evening, he rummaged through his basement and found some old tools. There was a logging stamp he wanted to bring with him, one of the old, iron stamps the Shaw company once used to mark its wood when the company still sent logs downriver. It was a double S design, as he recalled, and when he found his collection of stamps, there it was—a double S, just as he remembered it.

He found an old hand plane next that they could use to strip the bark. A broad ax? He might as well bring it. Who really knew what you needed to build this thing? And like that, looking for one old tool after another, Tom got ready for work.

As for me, I sat in my study printing off more pictures of rafts and looking for more information about Philemon Wright. Why in the world is there not more information about the man? All he did was settle the Ottawa Valley. There should be books about him. Statues.

Wright, I was surprised to read, never fully embraced the logging industry he created. Years before the Shiners—that brutal gang of lumbermen who tried to take over the river in the 1830's by assaulting and intimidating anyone who stood in their way—he despaired of life in the lumber camps and the proliferation of taverns in the Lowertown area of Bytown, that burgeoning community across the river from his home.

It was a far cry from the farmers' utopia he had envisioned. Even though the Wright family enterprises prospered, Wright devoted himself less and less to business as the years went on, and more and more to social activities, or the gardens around his stately home on what is now Taché Boulevard in Gatineau. He acquired the nickname the Squire. Helped establish St. James Anglican Church, the first church in the Ottawa valley.

And so it was left to the sons to control what their father had started, a near gold rush in the Ottawa Valley. Wright had nine children, and it is interesting to speculate on what might have happened with even a slightly different set of facts, for the future did not unfold as planned.

Wright's namesake, left at home to look after the settlement while his father piloted a load of square timber to Québec, died in 1821. Of Philemon Wright Jr., on whom Wright must have pinned great hopes and eldest-son expectations, we know little. He plays no part in the history of the Ottawa Valley, although he does hold the distinction of having one of the oldest marked graves in the region. Perhaps not what was intended.

Of Christopher Columbus, the youngest son, we know even less. The curiosity of the name makes you wish there were more of a story, but he died in 1841, a few years after his father, at the age of forty-two.

Ruggles made a brief name for himself—as though anything more were needed to bring that name into distinction—by designing the first timber slide in Upper Canada. He built it next to the Chaudière Falls, and although there were lumber companies who still tried to run the rapids to save money, or move in front of other rafts, the timber slide business proved profitable: most raftsmen saw the wisdom in a safe trip to the next stage of the journey, so they willingly paid the fare.

As for Tiberius, he would go on to have eleven children and face his own adversity. His first son, also given the name Tiberius, died before his first birthday. His first wife passed away a few years later. As if wondering whether his name was a jinx, he would wait thirty-four years before naming his last child, once again, Tiberius.

The young boy who piloted the first timber raft to Québec went into partnership with his brother Ruggles, and they ran the family business for many years. Still, it was Tiberius's second son, Alonzo, who would have the most success.

Alonzo Wright would become known as the "King of the Gatineau," the man who turned the family logging business into an empire; the man who became the first member of Parliament for the County of Ottawa after Confederation, and who would hold that seat for twenty-seven years until his death in 1894.

The second son of the second son. Who, in 1806, would have predicted any of that? You plan, you prepare—and then life begins.

Square timber by the shoreline in Braeside. (Photo courtesy of the Arnprior and McNab/Braeside Archives)

CHAPTER SIX

Deep River, Ontario, May 2008

I bump my way down another gravel road, until a large log home appears around a bend and I put on the brakes. There are cars parked in front. A knot of men standing around, talking. In the middle of the group, I see the Shaws.

I park next to the white pine that has been trucked over from the pole yard. The wood has taken up most of the front yard of the home, although there are other piles of uncut wood dotting the property. Some trailers as well. An old boat. Next to the boat—and it doesn't look as out of place as you might expect—a piano. It is a mad-hatter collection of odds and sods.

When I approach the group, a middle-aged man with cherub features is explaining how he is going to walk the wood through the saw, two other men helping; one side, then another, then the final two, and in that way the wood would be squared.

"Each stick could take a couple of hours," he is saying. "For the whole bunch, you're probably looking at two weeks. Longer, if Kevin here has to go look after his garlic field."

A long-haired, old-looking hippie shouts back:

"Longer still if I decide to go drinking."

"That's right," says the first man. "If we start drinking it's going to take longer."

Behind them, one of the meanest looking band saws I've ever seen stands ready to be turned on. Drinking? Not a chance. I'd want a bomb-removal suit just to get near the thing.

"We were hoping a week," says Dana Shaw.

"I was hoping to be a cabaret singer," says the first man. "We'll see what we can do. Come on guys, let's get to work."

And with that, I met Greg Merrill for the first time. I found out later he used to be a baker in Almonte, a job for which he would seem to have the perfect physical appearance. Perpetual grin. Fleshy features always moving in mirth. A belly he doesn't seem self-conscious about. You can picture him standing behind a pastry tray dressed in baker's whites. No leap of imagination required.

The bakery did well for a few years, then not so well for a few years, and in the late-nineties, he lit out for the territories, or in Merrill's case, Deep River, which passes nicely for the territories. He bought some land. Started building a log house. Bought a wonky old band saw to cut the wood for his home, then started doing piece work for the Shaws when they needed wood cut in Deep River. Before long, the baker found himself running a sawmill.

I stayed at Merrill's for most of the day, watching the first of the white pine get squared. The garlic farmer's name was Kevin Dupuis and he lived around Killaloe, had nearly two acres of garlic. Why one man needed that much garlic was a mystery to Merrill.

"No wonder you can't get a date," he said.

"We sell it," replied Dupuis. "I've told you that. We sell it."

"I've never seen you sell it. I've just seen you eat it."

As they talked, they moved the saw down the log, cutting off strips of bark, the air around them swirling with dust, so it looked like they were standing in the middle of a pollen storm.

"I've never heard you complain when I gave you some."

"That reminds me," said Merrill. "Don't give me anymore. I have enough garlic to end three marriages."

The saw dust twirled in the air. The bark fell to the ground. The garlic debate continued. Two hours later, our first piece of squared white pine was lying on the ground.

It was in those days in late May, driving up to Deep River, that I started to think about the raft in a different way.

If the maxim that states the sum is greater than the total of its parts ever needs to be proven, then build a timber raft. When you're done, calculate the weight. Take the dimensions. Crunch the numbers. And no matter what you end up with, you'll have a diminished picture of what you're actually staring at. The parts just won't add up.

I noticed it first with Johnny Shaw and his logging books. When he told me, yet again, that he had been reading the Charlotte Whitton book the night before, I started to realize he wasn't just looking for information. No one reads a Charlotte Whitton book that many times.

Ditto for Dana Shaw, who I noticed could not walk through the offices of Herb Shaw and Sons without stopping to stare at the black and white photos hanging on the walls. He had grown up around these photos, but he kept on stopping to stare, as if seeing them for the first time.

They had a lot riding on this, so I didn't think much about it at the time. Only after meeting Merrill, did the penny start to drop.

That first day, when not arguing about garlic, he talked about the heyday of logging. Talked about it as though it were last week.

"This is almost a king's log we have here," he would say as he measured the tree going on the saw. Then he told the story of how the British Crown, in order to keep the best wood for the navy, passed a law that gave it first rights to any tree seventeen-inches—or larger, in diameter. Pity the man trying to sneak wood like that into the harbour in Québec. Or a sawmill in the United States. It was the same offence as poaching in the King's forest, with the same possible penalty. Death.

"A king's log," Merrill would say again. "That's what we're working with boys."

Then he would tell stories about the timber rafts that drifted down the Ottawa River, and the communities that welcomed them. Long, colourful stories about Mattawa. Pembroke. Deep River (named because there are holes in the river bed here that go down 300 metres or more. "There are logs down there we're never seeing again boys.")

Greg Merrill singing the "Log Driver's Waltz." (Dana Shaw)

"Have you ever heard the 'Log Driver's Waltz'?" he asked me one day. I said I hadn't.

"Great song. I don't remember all of it. 'Birling down, down the white water,'" he started to sing. "That's part of the chorus I'm pretty sure."

When I was driving back to Ottawa that evening, it occurred to me that this was more than a job to Merrill. He cut wood all year long, so don't ask me when it happened, but this was clearly not just another job for him. The parts just didn't add up.

The difference between Dana Shaw's week and Greg Merrill's two weeks was split, and the pine was squared in a week and a half. On the day the job was being finished, I took another drive to Deep River, one of the final days of May. Along the way,

I noticed how the snow that so stubbornly clung to the shadowy parts of the forest was finally gone. And how the river held no ice, not even the fast-moving chunks caught in a current somewhere.

Summer might actually come this year. That seemed anything but a foregone conclusion until recently. Only a week ago, we had been felling trees in forests that still held snow. Been driving down gravel roads ringed with black ice. I still had a toque and pair of winter gloves sitting on the carseat next to me. Dana Shaw made a joke once that Tom Stephenson (who I still hadn't met 'cause he had been setting up shop at the pole yard) should design the raft to do double duty as an icebreaker.

Now, summer finally seemed around the corner. It was a pleasant drive up the Trans-Canada, noticing how everything around me was slowly starting to change, from trees to rivers, to the plastic window insulation I saw one homeowner removing from his house.

There was a noticeable change at Merrill's, as well. There, waiting for me were thirty former pine trees, now squared and cut into thirty-two-foot-long lengths. Although a lot of the weight and bark had been removed, in some ways it was a more impressive sight.

Merrill was standing by the logs waiting for me. His grin was larger than I had ever seen it.

"I found the song," he almost screamed to me.

"Which song?" I said, forgetting what we had been talking about earlier in the week. There had been a flurry of topics.

"'The Log Driver's Waltz,'" he answered. "Listen:"

And with that, the former baker from Almonte turned sawmill operator in Deep River, opened his mouth and sang:

"If you should ask any girl from the parish around
What pleases her most from her head to her toes
She'll say, "I'm not sure that it's business of yours
But I do like to waltz with a log driver.

"For he goes birling down a-down the white water
That's where the log driver learns to step lightly
It's birling down, a-down white water
A log driver's waltz pleases girls completely.

When the drive's nearly over, I like to go down
To see all the lads while they work on the river
I know that come evening they'll be in the town
And we all want to waltz with a log driver.

"For he goes birling down a-down the white water
That's where the log driver learns to step lightly
It's birling down, a-down white water
A log driver's waltz pleases girls completely.

To please both my parents I've had to give way
And dance with the doctors and merchants and lawyers
Their manners are fine but their feet are of clay
For there's none with the style of a log driver.

"For he goes birling down a-down the white water
That's where the log driver learns to step lightly
It's birling down, a-down white water
A log driver's waltz pleases girls completely.

"I've had my chances with all sorts of men
But none is so fine as my lad on the river
So when the drive's over, if he asks me again
I think I will marry my log driver.

"For he goes birling down a-down the white water
That's where the log driver learns to step lightly
It's birling down, a-down white water
A log driver's waltz pleases girls completely.

Like I said, something was happening.

CHAPTER SEVEN

Early June 2008

Here is a story about Tom Stephenson:
He was born in Perth, the oldest boy in a family of four, his dad being a mill worker for Tayside Woollen Mill, down by the Tay River, right in the heart of the town. When Tom was fifteen, not liking school much, he joined his dad at the mill.

He was a quick study. He worked himself up to the weaving room within a year. He had nimble fingers, a boy's enthusiasm, and at 6 foot 3 inches, he was hard to miss. The older workers told him to slow down, he was making them look bad.

He carried that gangly enthusiasm with him everywhere. To the baseball diamond. To the weekend dances, where girls noticed the good-looking, red-headed boy with the crazy two-step. He couldn't keep himself hidden even at the bank, where he went sometimes with his father. The bank manager took notice of the boy. The way he carried himself. The swagger when he walked through the doors.

The next time the manager saw Tom, he took him aside. Said he had a proposition for the boy. He had heard Tom knew some good fishing spots. Trout spots, in nearby Lanark Highlands, spots where a little creek you walk over every day might yield a prize-winning brook trout. Could they go fishing one day?

"I can pick you up this Saturday," the bank manager said, "if you think you can get me in there."

Tom looked at the bank manager and without giving it much thought said, Sure, he could do that.

"Be like your guide, right?" said Tom.

"Well, that and a bit more," said the manager. And then he walked away, using two heavy canes to move his body across the marble floor of the bank.

He had been wounded in the Second World War. Nearly lost his legs, although he was lucky — that's what the doctors said, he was lucky — and so he got to drag his legs behind him for the rest of his life, about as useless a pair of legs as a man could own.

That Saturday, the bank manager picked Tom up at his home and they drove to a lake outside Perth. There, after Tom got the manager out of the car, with all its fancy contraptions to let the man drive, he put him on his back and started walking.

They walked around the edge of the lake, then beside a small creek that emptied into it. The bank manager wasn't as heavy as Tom had feared, and it was easy to keep up a good pace. They walked beneath poplar trees just starting to bloom, then across open fields, back into the forest, the creek burbling beside them the whole time. That first day they caught their limit of fish, including a two-pounder that the bank manager almost refused to stop holding.

When deer hunting season came, they did the same thing, Tom carrying the bank manager on his back while they walked through the forest, looking for a good spot to set up a blind. Once they had chosen a spot, Tom would put the man down, brace his back against the trunk of a tree, put a Tayside Woollen Mills blanket over his knees, and there they would wait.

Sometimes Tom would leave the man while he walked through the forest looking for a deer he could direct his way, but for the most part they sat under a tree and talked. Tom told him how much he liked working at the mill, how the managers treated him with respect. He was starting to weave the samples the company sent to its suppliers, and there was no better job in the weaving room. Some men had worked at the mill for decades and never been asked.

They talked about their families. Work. The war. There were two generations between them, but for two years, Tom Stephenson took the bank manager from Perth fishing and hunting every Saturday morning, carrying the man on his back.

John Shaw, left, and Tom Stephenson look at plans in the pole yard. (Dana Shaw)

Then one day, as they were trout fishing, the bank manager told Tom he had a proposition for him. It would be up to Tom to accept what was being offered, although the older man thought he should. Would be a darn fool if he didn't.

"You can do a lot more than work at the mill, Tom," said the bank manager. "I've seen that with my own eyes. You need to go back to school."

"I don't like the school in Perth," answered Tom.

"I'm not talking about the school in Perth. You're too old for that anyway, Tom."

"What school are you talking about?"

"I'm talking about university, Tom. I know some people at the University of New Brunswick. I've talked to them about you. I think I can get you into the forestry program. Are you interested?"

A month later, Tom was on a train to the Maritimes, furiously reading books he had stuffed into a satchel, so he would be well-read by the time he got there.

I heard that story for the first time in the pole yard. It wasn't Tom Stephenson who told it, although he was there, nodding his head as the story was told.

"Yeah, that's pretty much how it happened," he said. "The bank manager's name was Mr. Robinson. A real decent sort. Not a bad fishermen, either."

As he talks, I stare at Stephenson with interest. No one is quite sure how old he is (and he won't tell), but he still has strands of red hair on his head and still looks down at most people when he is talking to them. He moves around in a herky-jerky, water-bug sort of way—long limbs in con-

stant motion, big strides around the stacks of wood—even if his shoulders are slightly stooped these days.

"Thirty tons?" says Stephenson, looking over at Johnny Shaw.

"'Bout that," answers Shaw. We are standing beside the square timber which was brought down from Deep River the day before. Tom nods and looks at the trees lying in the pole yard. Then back at Shaw.

"Thirty-two feet in length?"

"That's right."

"And to be accurate, it can't be more than twenty-eight feet in width. You know that, right?"

"To get down the timber slides. Yeah, I read that. I figure thirty trees for the base. Maybe less. Then whatever you need to actually keep it all together."

"You'll need some traverses. Probably two levels, I would think. Could be another dozen trees. And sweeps?"

"We'll need some way to row it."

"So you'll need sweeps. That could be fun. You'll need to carve them somehow."

Tom took out a small note pad and started scribbling into it.

"And what were you thinking of doing with the raft when you've built it?"

"We're not exactly sure. Put it in the river somewhere."

"To go where?"

"We're not sure about that either. We might need some sort of government approval to get down the river. We haven't looked into it yet."

"You're flying by the seat of your pants then?"

"Pretty much."

Tom nodded and continued walking around the trees, kicking at the frost on the bark, judging how many trees he would need to build the base of an old-fashioned timber crib. As he walked and did his mental math, he asked a last question.

"Why are you doing this, John?"

Shaw didn't answer right away. For a second it looked like he was doing his own mental math, then he said:

"I don't know. Maybe I just want to see one on the river again."

Tom didn't answer, just kept scribbling in his notepad, nodding his head, and I gathered from his silence that the answer was good enough for him.

When he put the notepad back in his pocket he said: "I'll get some tools out of my truck."

It's a nice way to spend a late-spring day, building a raft. There is something lazy and anticipatory about the work, much like the season itself. You're building something that will be put in the water in the summer. Not today. Not tomorrow. Just a pole yard worth of wood to play with, and no overly pressing deadline. You're working with hand tools. It's the second day of June.

The squared pine still needs work before it can be assembled into a raft. The saw has left a wane of bark on most logs. Using large metal scrapers that

Tom Stephenson and Algonquin College student Trevor Slack work on the raft in the pole yard. (Dana Shaw)

you hold in two hands, Stephenson starts to peel the last of the bark from the trees. Two forestry students from Alqonquin College— Trevor Slack and Isaac McEachern—work beside him.

The students have been conscripted by Fred Blackstein who teaches a course at the college. It's an easy credit at the end of the school year, peeling bark off white pine, working beside Tom Stephenson, who tells story after story and teases the students about not keeping up with him.

"It must be that Mary-Jane stuff you kids use," he says, as he takes a wide, deep cut down the log, peeling back the bark like it was a slab of meat, and he was some mad butcher. "Makes you all dozy and slow. Is that what you kids were up to last night?"

Protestations of innocence. Earnest attempts to match the monster cut that Stephenson just showed them. Two scrapers getting stuck in the wood at the same time. Under-the-breath curses.

"What's the problem? Did you go too deep?"

The students say, No, not a problem, what makes you think that, Tom? And right back to work.

It will take three full days to scrape the last of the bark from the pine. The work will all be done with the metal scrapers, Stephenson and the students bent over the wood like they were carvers working on a sculpture. Which, in a way, they are. Carving round wood to square, getting the angles right, lining it up, moving with the grain.

A loader with a mechanical arm on it will rotate the trees for the final cuts. In the heyday of rafting on the Ottawa River, it would have been teams of horses that rotated the wood and that moved the logs from one place to another. There must have been hundreds of teamsters in Eastern Ontario and Western Québec in the nineteenth century, a fact you might read somewhere and not think about, but when you see how much work goes into building just one raft, it hits home. Horses, and the men who knew how to work them, must have been the kings of any lumber camp.

I think of Willy O'Brien being the last full-time teamster in the Ottawa Valley and how, if he stopped working tomorrow, his labours would probably not be missed. No sawmills would sit idle. No lumber would fail to reach market. He could put the horses in the barn one night, not take them out the next day and that would be it. The end of an era.

What's more, no one would notice. Not a news story would be written. Not a television interview given. With the decision made by one man to not bother doing today what he has done his whole life, something real would become history. And the passing would go unreported.

I stared at Tom Stephenson carving away the bark on a log and realized, with a start, that we were trying to replicate that exact process, only in reverse.

CHAPTER EIGHT

Mid June 2008

The last square-timber rafts to sail down the Ottawa River were owned by J.R. Booth, and there is something appropriate about that because of all the accounts of the men, women, and families which flocked to the Ottawa Valley to seek their fortunes after Philemon Wright made his first raft trip, there is probably no better story than that of John Rudolphus Booth.

Born in Berlin (Waterloo), Ontario in 1827, he arrived in Wrightville in the early 1850's with nothing more than a grade-school education and a few carpentry skills. He was hired to help construct a new sawmill in Wrightville, and then for reasons that were never entirely clear, was given a one-year contract to manage the mill.

From there, he became an entrepreneur. He leased a shingle mill from Alonzo Wright, only to have the mill burn down the following year. This would be the first of many times when Booth had to start his business over again, virtually from scratch.

Around 1854, Booth moved to the other side of the Ottawa River, where he would live the rest of this life. He leased a sawmill on Chaudière Island and, in 1859, got his big break. The mill was awarded the contract to supply lumber and timber for the building of the Parliament Buildings.

The successful completion of the Parliament Buildings gave Booth a sterling reputation in the local business community. It also gave him access to capital which he used in 1867 to purchase the former timber limits around the Madawaska River owned by John Egan. Later, offered millions for the right to log on those lands, Booth turned down all offers and once said those timber limits were the basis for his fortune.

Ah yes, the fortune. Things moved quickly for Booth after 1867. He built sawmills up and down the Ottawa River, then in upper New York Sate and Vermont, even opened an office in Boston. He kept snapping up timber limits as well. At his height of power, Booth controlled 640,000 acres along the Ottawa watershed.

Between 1872 and 1892, Booth increased the output at his mills from 30 million board feet a year to 140 million. To put that into perspective, all the coastal mills in British Columbia at the time were producing 100 million board feet annually.

To get his wood to market, he started building railway lines, eventually owning the largest private railway in the world. To get his wood across the Great Lakes, he started a steamship company. By the end of the nineteenth century, it was rumoured he was the richest man in British North America.

Booth cut an imposing figure, even though he was short. He had a full white beard. Dressed in plain black clothing. Used the sort of language you never heard in the fine parlours around the Nation's Capital, and so he never became part of the social set. Which suited Booth fine. He would much rather be at his mills or out in his forests anyway.

Stories of how intimately Booth knew his business are plentiful (he was ninety-three before the business, a sole proprietorship, was finally incorporated.) One typical legend says he spotted a new horse in a lumber camp on the day it arrived, even though the camp had more than 500 horses. Other stories talk of him knowing the names of every logger who worked for him. Or how he would hire clerks by leaving a piece of paper on the floor and hiring the first job applicant who thought to pick it up. In 1920, when he dropped the puck at a Stanley Cup hockey game between the Ottawa Senators and the Seattle Metropolitans, he received a five-minute standing ovation.

When he died in 1925, at the age of ninety-eight, Prime Minister William Lyon MacKenzie King declared a day of mourning and called Booth "one of the fathers of Canada." He was buried next to his wife at Beechwood Cemetery in Ottawa.

In the story that ran in the Ottawa Journal on the day of the funeral, Michael Gratton O'Leary said Booth should not be remembered "as the great magnate whose wealth is the envy of many and the wonder of more; but the great pioneer, the man whose genius and imagination tamed the wilderness . . . and, above all, did more than many of his time to build up this Ottawa Valley."

In 1908, working a timber limit off the Coulonge River, Booth lumbermen built 150 cribs of white pine for market in Québec City. Raft traffic on the Ottawa River had slowed to a trickle in recent years, and news of the cribs being built soon spread up and down the Valley.

There were other reasons for the excitement. That year, work had started on a hydro-electric dam at the Chaudière Falls. When it was completed that autumn, it would raise the level of the river ten feet behind the falls and make the timber slide useless. There would simply be no way for a raft to get past Ottawa after the summer of 1908. Booth's 150 cribs would be the last to make the trip.

The loggers toiled away all winter felling trees, squaring them with their broad-axes, then building rafts on the frozen Coulonge River. When the river melted it was like an armada had just been launched, so many rafts travelling together they stretched to the horizon and beyond.

By the time the raftsmen reached Arnprior, there were crowds waiting at the wharf to see them. Somewhere among the crush of people was a young man by the name of Fred Stickler.

Rowena Stickler takes the photo album from her purse and places it on the table.

"Uncle Fred always talked about it," she says. "Ever since I was a little girl, I remember him telling stories. The last raft trip down the Ottawa River."

Stickler starts to turn the pages of the photo album. Black and white photos flick by my eyes—old houses, men in suspenders, women in summer frocks—as she continues talking:

"He was only eighteen, and I don't know if he went to Arnprior to try and get on the rafts, or whether it was a spur of the moment thing, but Uncle Fred was real proud of being on that raft. He told me the story many times."

Her uncle, she says, never married. He spent his life living on a farm near Renfrew, one of those old bachelor farmers you run across in the Valley from time to time. A little gruff. A little squirrely. For most people who knew him, other than a young niece who adored him, he came across as a pained eccentric, almost a hermit.

But when he was a young man, he was full of bluster, and he talked his way onto the last raft to travel the Ottawa River. According to his niece, he was hired as a casual labourer and accompanied the rafts as far as Montréal. He spent a night in Ottawa, drinking at taverns in Lowertown, rode the chutes down the Chaudière Falls, and had stories to tell for the rest of the summer.

She points to pictures of her uncle in the photo album. One shows a young man with slicked down hair and a stiff white collar. He has sharp, angular features, is handsome, although he does not smile at the camera and looks a bit sad because of it. The other photo is of an old man with white hair and worn suspenders standing in front of a little cabin with long grass growing around it. Fred Stickler, with a few decades behind him.

"I think Uncle Fred would have liked being a raftsman," says his niece, pausing for a minute from flipping pages of her photo album. "I think that life would have suited him to a 'T.' There was always something about Uncle Fred that made you think he wanted to be someplace else."

I look at the sad-looking cabin where he eked out a living and am not surprised. The elderly woman in front of me goes back to showing photos in her album. Rough-looking men—these could be photos of the Hole-in-the-Wall Gang—stare defiantly at the camera. Women with white summer dresses occasionally pose with the men. A few photos are professionally done, the same people dressed in stiff clothing, with the name of a photo shop in Renfrew on the back.

Rowena Stickler contacted me after hearing about the raft. Drove out with her photo album to see it. Wanted to see the "boat" that was the subject of so many childhood stories told back in the forties while she sat on the knee of her bachelor uncle. Stories of raftsmen and river trips and a time before the Great War when the rivers in the Ottawa Valley would be teeming with Huck Finn-crafts all heading east.

"Uncle Fred was only on a raft once, but it was important to him," says Rowena Stickler. "I've often wondered why that was."

It might have been the biggest day of his life. That might have had something to do with it. When the raftsmen reached Ottawa that summer of 1908, there were so many people waiting to see them the police had to be called out to control the crowds. Newspaper stories were written about the J.R. Booth Armada and its swan-song journey to Québec. Considering that raft trips were once as common in the Nation's Capital as snow, this might have been a first.

The crowds continued coming out along the river after the raftsmen left Ottawa. In Hawkesbury, Pointe Fortune, even in Montréal and Québec, when the rafts hit the St. Lawrence River. Thousands of people, who had grown up with the raftsmen and the annual river run to Québec, came to see the last raft from the Ottawa Valley.

Top left: *Fred Stickler, at roughly the age he would have been when he jumped aboard the last square-timber raft, in 1908.*
Top right: *Fred Stickler, left, with family members outside his home in Renfrew County.*
Bottom: *Fred Stickler, second from left, at a logging camp in Renfrew County. (Courtesy of Rowena Stickler)*

And then, just as quickly as the crowds arrived, the story that happened in this place that was still a frontier in many ways, was forgotten. A Great War, the recession that preceded it, the daily trials and tribulations of surviving in Upper Canada—which a lot of the old timers still called it— followed. So perhaps the forgetting was not surprising. People had other things on their minds. No one had the time to remember a raft.

Until 100 years had passed, and some lumbermen up the line decided it was worth remembering.

A cookery crib. (Photo courtesy of the Arnprior and McNab/Braeside Archives)

CHAPTER NINE

Pembroke, Ontario, June 2008

When the white pine is squared, scraped, and bevelled at each end, Tom Stephenson decides the wood is finally ready for building a raft. He gets the loader with the mechanical arm to place a couple of hydro poles at the far corner of the yard, then gets the driver to start placing the squared timber lengthwise on top of the poles. Again, I am reminded that all this work would have been done, at one time, with horses and teamsters moving these giant trees around the surface of a frozen lake.

Even though we are cheating, it still takes several hours for the timbers to be placed where Stephenson wants them.

"I think the biggest pieces should be used for the traverses on top," he yells at the truck driver. "We could use them for the outside beams, but I think on top is probably how they did it."

The driver places another log across the hydro poles. Then he nudges it, with the back-side of the mechanical arm, snug against the log he has already placed on the poles. The biggest logs measure more than a foot-and-a-half in diameter, and although we have trimmed a lot of wood away, the total weight of the raft will still approach thirty tons. It's a good thing we have a loader and a driver who's skilled enough to pat the timber into place.

Stephenson starts to walk over the assembled beams. He struts his way down the logs like a bantam rooster.

"That's right," he yells. "On top. Just put it right on top here. "

The truck driver is at the far end of the yard, getting another piece of squared pine. I look around to see whom Stephenson is yelling at, but I don't see anyone.

"Right on top should do it. What do you think? Starting to take shape, eh?"

Stephenson has his back turned to me. I do a full 360 trying to see if someone has entered the pole yard, but I don't see a soul. After a couple of minutes of animated, shouted conversation, I realize Stephenson doesn't need an actual person to carry on a conversation.

At lunch, when the base of our raft has been laid down, I ask him about it, as subtly as you can ask a question like: "So, how long have you been talking to yourself?"

And he answers:

"It must come from working in the bush. I just start talking. Hell, if I needed someone to listen to me, I'd be shut-up half my life."

By the end of the first week of June, the bare-bones structure of a timber raft have begun to appear in the pole yard of Herb Shaw and Sons.

Stephenson has used twenty-seven white pine in total. He has spread twenty out for the base, then used three half-split pieces for the traverses. On top of the traverses, laid perpendicular, were four more squared pieces of pine. The pieces on top were the largest.

To keep everything together he has not used a single nail, screw, dovetail, notch or strand of twine. Nor would the lumbermen of a century ago. Figuring out how they actually pulled off a stunt like that was one of the trickiest parts of designing the raft.

What Stephenson devised was a simple pin design using the diagrams from the Charlotte Whitton book, and if you look at the old photos, you sure enough can see pins on most of the rafts. To do this, holes were drilled along the rectangular base of the raft. The holes went through the traverses and a bottom log, but not through the four monster logs that sat on top. Those pieces fit between the protruding pins, and the weight helped keep everything together.

Stephenson used ironwood for the pins, and late one afternoon as he was carving one out, he marvelled at the design that Philemon Wright probably came up with.

"The logs in the middle roll free," he said. "That's what makes it work when it's going down rapids and chutes, otherwise it would crack up. It would be too firm.

"The outside structure keeps it all in place. Just the pins and the traverses, that's what makes it work. It's sort of like a log boom, the wood cribbed together in the middle. That's the word he must

have used. Cribbed. That's why people started calling them cribs."

The raft is much different from what would have been built along, say, the Mississippi River of Mark Twain. Along that slow-moving waterway, rafts could be one level, could be twined together, and had no need for extra weight or flexibility. Nor were they the narrow rafts of interior British Columbia that were rarely tied together, the rivers there being too narrow.

The rafts of the Ottawa Valley were large, rugged and perfectly suited for the task of running rapids along la Grande Rivière du Nord. How Wright could have thought of such a vessel, coming from the placid waters of New England, is a marvel.

Our raft was starting to take shape, and right around the same time, the strangest thing happened. Visitors started to arrive in the pole yard.

They came from around Pembroke at first. A timber broker who dealt with the Shaws. A forestry teacher at Algonquin College. A friend of Greg Merrill's, just passing through.

The pickup trucks would pull into the pole yard and men would get out, walk up to the raft, and almost to a person the initial reaction would be the same—a ball cap would be pushed back on a head, a temple would be scratched, there would be a low whistle, and then a comment something along the lines of:

"Didn't think it would look this big."

Or:

"Looks different from the photos. Is that ironwood you're using for the pins Tom?"

That's another thing I noticed: everyone who arrived in the pole yard seemed to know Tom

Stephenson. Although every second person called him a different name.

"How ya doin' Red?" someone would say.

Or:

"Hey, Soupy, is that you?"

Stephenson, carving out a pin from a stick of ironwood, would answer: "Not bad. Doin' well. What do you think of my raft?"

I asked him one day how many nicknames he had and he shrugged, flashed that smile he has—his wife Judy says her husband smiles so often "it's eerie,"—and then he said: "I've probably got nicknames I don't know about. Working in the bush, you get called a lot of names."

Then one day a woman arrived from the *Pembroke Observer* newspaper, and a few days after that people arrived from all over the place. Mattawa. Golden Lake. Eganville. Most said they were just passing through, one place to another and all that, decided to "swing by" and see the raft. More than one person, however, admitted they had made the trip to Shaws for no other reason than to see the raft. A return trip—for some—of more than 300 kilometres.

Again, I was struck with the sense of something happening here that I didn't fully understand. Like Greg Merrill, a man who works sawing wood all day long, suddenly learning the "Log Drivers' Waltz." Or Johnny Shaw staying up late, reading logging books.

Whatever was happening, by the time the last days of spring had begun to fade, and a new season had dawned with little apparent enthusiasm— the days were grey and rainy, with little warmth even at mid-afternoon—there was a steady stream of traffic arriving in the pole yard.

I watched grown men walk across the base of the raft and start to laugh. Others run their hands down the smoothly shaved timber. Pickup truck after pickup truck.

There was still the little problem of putting the raft in the water. In the rush to get the vessel built, no one had got round to contacting Transport Canada or the National Capital Commission or any of the other myriad government agencies that might need to be reached for approval to take an old-fashioned timber raft down the Ottawa River. Another meeting was convened in the offices of Herb Shaw and Sons.

Dave Lemkay starts the meeting by suggesting we put the raft into the Ottawa River at Petawawa and take it as far as Pembroke.

"We can be in and out in a day," he explains. "We might be able to avoid a lot of government red tape that way."

A good plan. On several points. Although there does seem to be one noticeable drawback, and Dana Shaw is the one to articulate it after an awkward pause in the room.

"Shit, Dave, that's not much of a raft trip."

"Well, it would do the job," he answers. "We can put it in the water, show everyone it floats, take some pictures and we've done it."

"Sure," says Shaw, and for a few seconds you can see him struggling for the words to articulate the opposite school of thought. Then he says, "But don't we want to play with this thing a little longer?"

Lemkay asks what he is considering. Shaw says he has no idea. People nod their heads. No idea.

Before long, a map is found somewhere and spread on a table in front of us. The number of dams between Pembroke and Ottawa are counted. Three. That's a problem. Did anyone not realize there were three dams? Someone says, "I thought you knew."

"Do we want to take it all the way to Ottawa?" asks Johnny Shaw. This is briefly debated. Someone suggests taking the raft all the way to Québec. Someone else suggests putting it in a lake somewhere first just to make sure it floats, and we don't make fools of ourselves. The lake idea is debated and then discarded. Along the way, we forget to answer the question about going to Ottawa.

"Cost is another factor," says Johnny Shaw. "This is starting to add up. We have about $30,000 in wood tied up in that raft. Now if we have to start trucking it around the dams, well, I don't know how much this will cost before we're done."

"Maybe we can get some sponsors," says Lemkay.

Everyone loves the idea of getting a sponsor. No one knows quite how to do it.

"What would they be sponsoring?" asks Dana Shaw. Someone suggests individual logs. Someone else suggests a leg of the trip.

Back and forth it goes. By the end of the meeting, very little has been decided other than I will try to find a sponsor in Ottawa, and Dana will try to figure out a river route. As we walk away it occurs to me we forgot to talk about government approvals. Maybe next meeting. We really have to start taking minutes.

There are finishing touches. I arrive in the pole yard one morning to find Stephenson carving a ten-foot oar, a "sweep" as the raftsmen called them. It's ridiculously over-sized. I put it in my hand and feel immediately like the Jolly Green Giant.

I stay for the day, and there is a period, almost an hour, when no strangers arrive in the pole yard. In that time I sit and watch Stephenson carve his oar. He is a skilled carver, makes wooden marionettes at home, has dozens of the puppets in his eclectic collection (he often carves political figures and had puppets of Maxime Bernier and Julie Couilliard until, after their five minutes of fame, the puppets were mercifully retired). Even the briefcase he uses is wood, with wooden handles he has carved.

"I love wood," he says. "You can do anything with it."

His job in the lumber camps for many years was grading the wood as it came out of the bush. He could look at a tree before it was felled and tell you how many board feet you would get from it, and what grade the wood would be when it came out of the mill. Knew which pine would end up being used for veneer, and which would be used for lumber.

He even made a display once of white pine as it would have been graded at the lumber camps, then gave it to the curator of the Champlain Trail Museum in Pembroke. You could see clearly, in the fourteen samples, the knots, the grain, the way the wood looked when it was burnished. He thought it was a lovely display and didn't understand why it wasn't front and centre at the museum.

For nearly an hour, he talked about wood. In that time, talking about a student he once had who could never grade pine to save his soul, I tweaked to the fact he had once been an instructor at Algonquin College.

"You were a teacher?" I said.

"When the children were growing up," he answered. "I didn't want to be in the bush all year long when that was happening, so I took a job at the college."

"For how long?"

"'Bout twelve years. 'Till the kids were old enough. Then I ran back to the bush 'bout as fast as I could run."

"That was a good job to give up. Why did you do it?"

"Good job?" he answered. "Hell, I couldn't wait to quit. Stuck inside all year long. And for what? Money? Why would anyone do a thing like that?"

He told me this story on that quiet afternoon while we sat in the pole yard, me sitting there watching him carve an oar, talking about wood, and I'm here to tell you it didn't strike me as a strange comment at the time. Just seems strange when you type it later.

On Thursday, June 19, I received a phone call from Dana Shaw.

"We're putting the raft in the river next Monday."

Tom Stephenson walks toward the square timber crib in the Herb Shaw and Sons Limited's pole yard in Pembroke on June 5, 2008. (Darren Calabrese/ Courtesy of the Ottawa Sun*)*

"Where?"

"In the Bonnechere, where it hooks up with the Ottawa River at Castleford."

"Are we going all the way to Ottawa?"

"That's the plan."

"Great. What time are we leaving at?"

"First thing in the morning. We'll be there around six. Dave Lemkay will be there with a sweep boat as well, in case we need it."

"Who's going on the raft?"

"You and Darren (Calabrese, a photographer with the *Sun*). Me. A couple of other people. And Tom."

"Tom's coming as well?"

Shaw laughs.

"Try and keep him off that raft. I don't think you could."

Tom Stephenson brought the canvass tarp from the basement and unfurled it. The weather was calling for rain tomorrow. They would need some sort of shelter. He carefully started to inspect the tarp for holes.

He had already cut some poles. He figured the tarp would need six, maybe one more for the centre. He was thinking of flying some flags as well — the Maple Leaf, the fleur-de-lis, the Ontario ensign — and if he was going to do that he would need to cut some more ironwood branches. He thought the flags were a good idea. He would need to find the time.

His fingers ran over the harsh canvass looking for weakened seams, holes, frayed material. It had been a long time since he had used this tarp. Never thought when he packed it away in his basement that he would be bringing it out again for use on a raft.

It had been a fun month in the pole yard. He had enjoyed the work, planing the bark off the trees, cutting the tapered ends with a chainsaw, full days spent working with wood in what might as well have been the biggest wood shop he had ever been in. Even the smell of the pole yard — the pine, the sawdust — it was wonderful. He awoke in the morning anxious to leave his bed and get back to work.

The tarp looked fine, and he rolled it back up. He went back to the basement to get the flags, then started loading his truck. He would leave a little earlier tomorrow and stop at a hardwood stand he knew on the way to the Bonnechere where some ironwood was growing. He would get the flag poles there.

He went and got his rain gear, put that in the cab of the truck as well. Then his chainsaw, a toolbox, the poles he had already cut. Thirty minutes later the truck was packed and Stephenson headed upstairs to his bedroom. He took off his shirt and washed up. The weather forecast was for rain for most of the week. It didn't seem right, after the winter they just had, that summer wasn't a bit more obliging. But what could you do? And besides, they wouldn't be the first people to get wet on a raft. Look at some of the old photos of a raft making its way through a set of rapids somewhere on the Ottawa River, and you shouldn't complain about a bit of rain.

He crawled into bed and reached over to set the alarm. The Shaws wanted everyone at the wharf at 6 a.m. He did some mental math, taking into account the flag poles he needed to cut down, and set the alarm for four. Then he rolled over, rolled over again, back and forth like that, until somewhere in the far corner of the room he heard a screeching sound, and after a few minutes he turned off the alarm. Before leaving the house he remembered to grab a fishing rod and tackle box.

Dana Shaw took the moose burgers wrapped in butcher's paper and placed them tenderly in the cooler. He placed tomatoes, lettuce, carrots, peanuts and Export beer on top of the burgers. Then a bag of ice. In a dry-goods crate, he placed hamburger buns, potato chips, plates and cutlery. He opened the cooler and had a last look, then did the same with the crate. It looked like enough food.

They would restock every day anyway. Johnny would drive down with more food for them, so he wasn't that worried. He had sausages and steak already picked out, some salads as well. He figured cooking on a barbecue would be a little bit like the old campfires on the rafts. They probably ate a lot of moose back then as well.

Earlier that day, he and Tom had supervised the dismantling of the raft, numbering the logs as they took them apart so they would know how to put it back together again, then placing them on the logging truck for the drive to the Bonnechere. They hadn't bothered to weigh the wood, but Tom figured it was close to thirty tons. Dana thought it might be a bit more.

It had felt good working with Tom this past month. Dana had enjoyed leaving his paperwork on his desk and driving down to the pole yard early in the morning, talking to Tom. Enjoyed watching the raft take shape, the smell of the white pine as it was cut into shape, even enjoyed the ache in his arms after he had planed some of the bark away.

"Hey, lumberman," Tom would shout whenever he arrived. "What do you think of our raft today?"

Lumberman. That's what he and Johnny had been called their whole lives. The Shaw cousins. Fifth generation. Lumbermen going all the way back to 1847, and while the nickname bothered him when he was a teenager—like all teenagers he wanted to be seen as something other than the shadow of his family—he didn't mind the term now.

"Lumberman," he would shout back at Stephenson. "I think she's coming along. Did somebody teach you how to do this last night?"

And in that way, they would spend an hour or two every morning working on the raft, trading jokes, until guilt over the papers left on his desk would finally drive him back to the office.

There would be no more guilt and no more interruptions, come tomorrow. The work at Herb Shaw and Sons would be left for Johnny, while he and Tom piloted the raft to Ottawa. The reporter and a photographer would be on board as well. Along with some friends. Seven people altogether.

He took one last look in the cooler, debated whether to throw in more beer, decided he should, then left the cooler by the front door. He went upstairs and got ready for bed.

I had my own last minute errands to do. I had promised to try and find a sponsor for the trip.

For several days I racked my brain, trying to think of whom I could approach. It needed to be someone who loved history, who might see the crazy charm in this project and want to get involved. Someone who could cough up money right away.

The first criteria seemed fairly easy. The second one eliminated just about everyone I thought of.

No corporation, museum or NGO could make a decision that quickly. I would be right back talking to people suggesting we try this next year.

I thought about it, made lists, thought about it some more and then, finally, it came to me. I jumped in my car and drove to Minute Car Wash on Catherine Street.

Tony Shahrasebi—or Tony Q as most people call him—was right where I thought I would find him, behind the car wash with a shammy in his hand. He owns the car wash—he also owns Capital Parking and several rental properties in the city—but I never even knew he had an office until I dropped by that day.

"Tony, do you have a couple of minutes?" I said when I arrived.

"Sure, sure, come," and we walked to that office I'd never seen before. Tony is an Iranian immigrant who came to Canada after the fall of the Shah, and he went on to make a fortune in the parking business in Montréal. He became a millionaire after he came up with the idea (you might want to sit down for this) of "tandem parking."

Tandem parking is when you park in a lot and leave your keys with the attendant. The attendant then squeezes every possible car he can onto the lot, sometimes tripling the number of cars that can park there.

Tandem parking made Tony Q a rich man (and you thought software design was the way to go) and in the mid-nineties, he bought Capital Parking and moved to Ottawa. I had done stories on Tony Q and knew he was a history buff who often complained that people in Ottawa had no appreciation for it.

"I come from a country where something has to be more than 1,000 years old before it is considered historical," he told me once. "In Ottawa, nothing but the river is 1,000 years old. And it seems people want to keep it that way."

We walk to his office which I would have bet money had little pretense or frills, and when we walked into a room that looked like the waiting area at a Canadian Tire garage, I was not disappointed.

"So what's up?" he said as we sat down.

I told him about the raft. Then about J.R. Booth. Philemon Wright. And finally, these crazy people in Pembroke who had built an exact replica of a square-timber raft and now wanted to bring it to Ottawa.

"They're looking for sponsors," I said. "I thought of you."

"It's a raft?"

"That's right."

"Like Huck Finn?"

"A bit more than that," I said, and took some photos out of a folder. I showed him the raft as it sat in the pole yard earlier in the week, almost completed except for the sweeps. Then I showed him archive photos—rafts tied up below the Parliament Buildings, a solitary raft making its way down a treacherous set of rapids.

"This is how they used to get the wood to market?" he says.

"That's right."

"I had no idea."

I show him more photos. Tom Stephenson planing the bark off a square timber. Willy O'Brien taking down a tree in a forest near Deep River. Another archive shot of rafts being built in a lumber camp somewhere. Tony starts to laugh.

"Man, this is all new to me. It looks crazy."

"It is crazy. Do you want to be a part of it?"

He keeps laughing as he opens a drawer in his desk and takes out a cheque book.

"What am I actually sponsoring?"

"We're not sure."

"What do I get for sponsoring this raft?"

"We're not sure about that either."

The laughter now can be heard out in the car wash.

"All right, how about this one: Who do I make the cheque out to?"

God love history buffs. We're on our way.

And off we go. A square timber raft about to go down rapids on the St. Lawrence River.

PART TWO

Dignitaries go down the timber slide at the Chaudière Falls.

CHAPTER TEN

Castleford, Ontario, June 23, 2008

At 6 a.m., the raft—disassembled and sitting on a truck—arrived at Horton Township Wharf on the shores of the Bonnechere River in the heart of Castleford, a community founded by a mid-nineteenth century English naval lieutenant turned wannabe lumberman, right where one of the Ottawa's logging river tributaries enters the main "highway." There, it was reassembled, using a numbering system Stephenson had come up with, the numbers branded on the end of every log that took a pin.

The mouth of the Bonnechere River is a narrow strip of water with overhanging willows and dense bush on the east shore. When dawn came there was no sun, only a grey sky that was already turning black. Headlights were left on, so you could better see the timbers as they were hoisted off the truck with a mechanical claw, the wood flying through the night air until it gently landed in the water.

It was a strange sight. But perhaps a stranger sight was the parking lot of the wharf itself, where pickup trucks and mini-vans were parked. Far too many for the number of people who would actually be travelling on the raft.

People had come to the wharf to see the raft being put together. A story had been published in the *Ottawa Sun* that morning—the first story to be published after nearly two months of driving back and forth to Pembroke—and I walked around the parking lot, surprised to be interviewing people.

"Never thought I'd see one of these," said Bill Thompson, a retired plumber living in Ottawa. "My dad used to tell me stories about these rafts when they came down the Ottawa River. So many of them you could walk across the river. That's what he used to say."

Thompson stood in a knot of other people, craning their necks to see through the darkness. He had driven up from Ottawa. Would travel down to Braeside to see the raft when it pulled in for the night.

"Never thought I'd see one," he said again. "I don't know how they managed to build it."

I looked around and saw Tom Stephenson walking over the base of the raft, yelling instructions to no one in particular. There's your answer, I almost said, but instead, I nodded and walked away to the next group of people.

For nearly an hour I spoke to people, amazed so many had turned out. I had hoped the raft stories would be interesting, that if I were lucky some readers might look forward to the Sunday paper and the

*Dana Shaw, right, and Tom Stephenson row the timber crib along a portion of the Ottawa River before
landing just upstream from Gillies Old Mill in Braeside Monday, June 23, 2008.
(Darren Calabrese / Courtesty of the Ottawa Sun)*

next installment. But to get out of bed before dawn, drive in the dark to Horton Township Wharf, I wasn't expecting that.

It took more than two hours for the raft to go back together again, and in that time no one left. Much of the work was done in the dark with a battery of headlights turned towards the river as the giant logs moved in and out of the arc of light. If there were a stranger sight on the river that morning, I have no idea what it might have been.

Shortly before nine, the last of the coolers and rain gear were loaded onto the raft. Dana Shaw's barbecue, which became a symbol of the trip, was placed carefully on one of the top beams. The oars were placed in their locks. And like that, with an unexpected crowd to wish us Bon Voyage, we are on our way.

We row our way from the wharf, but quickly take advantage of a sweep boat that Dave Lemkay has found, borrowed and brought to the wharf along with startled owners, Peter and Barbara Haughton.

In the century since the last raft travelled down the Ottawa River or any of its tributaries, dozens of dams have been built. Every river has seen its level rise. Every river has been flattened and calmed, some to the point of being little more, today, than meandering bathtubs.

There is simply no current on the Bonnechere, and we hook up with the sweep boat so it can tow us out to the Ottawa River. Overhead, a few raindrops are falling. I pull on a raincoat and look around.

It is like a canal cut along the Rideau River, the way the Bonnechere looks at its mouth. The trees overhang the banks. The river is straight and narrow, then abruptly ends in a small marsh along the shoreline of the Ottawa. Looking over at this tributary from the main channel of the river, you could scarcely guess that it is a major river itself, tumbling out of the Algonquin Highlands far to the west and filling both Golden Lake and Round Lake before it reaches this little marsh.

As for the Ottawa River, somewhat different. We are on a stretch of the river so wide, it is called Chats Lake on some maps. There are islands. A point of land on the far side so large it is almost an isthmus. But the current is no stronger.

We head east, letting the sweep boat—it is of fifties vintage, once used by the Gillies Lumber Company to move log booms around—tow us through the rain. The engine of the boat sputters and misfires from time to time, belching out black smoke that disappears quickly in the rain.

I look around the vessel and see the faces I will spend the next few days with: Dana Shaw—it is not yet ten in the morning, and he is already fiddling with his barbecue; Paddy Irving, a mill hand at Shaws; Jean Gougeon ("Gouge"), a friend of Dana's who got married the week before and showed up at the mouth of the Bonnechere after Dana asked

if the honeymoon was over, and did he have a few days to spare? He arrived not even knowing what was happening, staring in wonder as the timber was loaded off the truck in the predawn darkness. Darren Calabrese was there, running around the raft trying to keep his cameras dry.

And Tom Stephenson, sitting atop one of the beams, sporting a goofy, Jethro-Bodine hat and smiling to beat the band. When I arrived at the wharf, I was pleased to see a tarp overhang pitched on the raft, something I was guessing Stephenson had concocted the day before when he heard the weather forecast. But the man was making no attempt to huddle under it, like the rest of us.

"Rain won't hurt you," he teased. "Makes it seem more like a raft trip, if you ask me."

And there he rocked on the back beam, sitting cross-legged and smiling at us. We left him there as Darren and I started examining the floor of the raft. Others were looking at it as well, although we were the only ones being quite so obvious about it.

For many weeks now it had been hotly discussed how the base of the raft might function. Sure, it would float. But if the wood was not squared properly, if the base was off kilter, and then if there was any sort of wake working against you, exactly how dry would you stay?

Oh, the theoretical arguments. They abounded. But now the experiment was underway, and everyone looked at the wood beneath him with more than passing interest. There was a definite backwash at the front of the raft where it cut into the river. Standing up there it felt like you were surfing. But other than that, and the few times when the wake from a passing boat was large enough to roll over the timbers, it seemed dry enough.

In a few minutes, my muscles relaxed. I ventured out from under the tarp. Looked around. It is a fat, lazy stretch of river this, and we putter along the Ontario side. Giant white pine can be seen on the shore, their fine, effervescent needles obvious in the treeline. Granite cliffs drop into the water in a straight line, like some giant step from Big Joe Muffraw.

There are parts of the Ottawa River further north, that have changed little from the days when Samuel de Champlain travelled upriver from Québec. Wild stretches of water that have not been touched by hydro-electric dams or settlement. The river here is far too calm for that, but the sense of isolation from time to time is perhaps the same. There are stretches here that are nothing but treeline and granite escarpment.

We are huddled under the tarp, with the exception of Tom, when it suddenly occurs to me—did we ever contact any government agency about putting the raft in the river?

I ask Dana.

"I don't know," he answers. "You should ask Dave. I think he was looking after that."

I nod my head. No one has asked. It's as obvious as the rain falling on my face.

I've been on the Ottawa River often enough to realize that OPP cruisers are not a rare sight. And last time I had the opportunity to converse with the water police, I came away with the distinct impression they were rather keen on all these red-light-green-light-lifejacket-bailing-can-whistle rules they have nowadays.

This could be a short trip. I go to the back of the raft and sit beside Stephenson.

"Don't worry about it," he says. "Dana knows a lot of people. Plus Gouge has a friend joining us tomorrow."

"I don't know," I say. "Maybe we should have told someone we were putting this thing in the river. If we get stopped after already publishing the story in the *Sun*, it's going to be embarrassing."

"I think Gouge's friend will be able to handle things," says Stephenson. "We just have to make sure we don't get into any trouble today."

"What's so special about Gouge's friend?"

"He's an RCMP officer in British Columbia."

"So?"

Stephenson looked at me as though I were new to this river rafting game. As though it were obvious.

"He's got a badge. He's a fed. No OPP officer can bother him."

So we're travelling down the Ottawa River. Trying not to draw attention to ourselves. As inconspicuous as a thirty-ton raft being towed by a sweep boat belching black smoke can be. Not to worry. If anyone stops us, assuming there is no problem today, we will have a holidaying RCMP officer aboard.

"Do you think this is going to work?" I say to Shaw as he passes me a moose burger right off the grill.

"We're going to find out," he answers. I notice that's his answer to a great many things.

CHAPTER ELEVEN

June 23, 2008

I huddle under the tarp, thinking of raft trips I have read about in the last few days. There was the last trip down the Ottawa River in 1908, the one we are honouring by going down the river this year.

There was Philemon Wright's trip in 1806, the one which opened up the Ottawa Valley to logging. I kept searching the Internet, looking for more stories. Finally found one about a raft trip made on the Susquehanna River in Pennsylvania in 1938.

The forests of Pennsylvania and the nearby Ohio Valley once rivalled the Ottawa Valley for their great stands of white pine and its many fast-charging rivers. Timber rafts were a common sight along the Delaware, Susquehanna and Schuylkill rivers throughout the nineteenth century.

And then, just as happened along the Ottawa River, the timber rafts disappeared. Dams were built. The rivers were raised. Railway lines appeared nearby. Ships stopped being built with wood. Just like that, as though fingers were being snapped, a perfect storm of different facts came together, and the raftsmen faded away.

Then in 1938, in the final days of the Great Depression, some of the men who travelled on those last rafts came up with the idea of building another one and going down the river one last time. Might as

well, the argument might have gone. There wasn't a lot of paying work to occupy their days.

The journey, in a surprise twist, caught the fancy of the American public. Movie newsreels were shot of the journey and Harry Conner, the seventy-five-year-old pilot of the 112-foot, 80-ton raft became a media celebrity. As he successfully navigated the raft past the dangers of Rocky Bend, Chest Fall and the dam at Lock Haven, the story grew and grew.

Newspapers in the mid-west started to cover the journey as a serial. Reporters from New York — the *Sun*, the *Herald Tribune* — descended on the towns along the Susquehanna River, and for decades afterwards people in those sleepy river communities talked of the strange, hard-drinking, tip-crazy Yankees who came to their towns in the spring of 1938.

The plan was to pilot the rafts to a mill in Fort Hunter, just above Williamsport, where the wood would be sold, like in the good old days. Every aging lumberjack would leave with a bit of coin in their jeans. That would have been part of the charm of the serial in 1938.

And then near Muncy, Pennsylvania, in a twist no one could have seen coming — although in hindsight the smashing of one era against another does not seem implausible — the raft ran into a railway overpass. There, it promptly broke apart.

Harry Conner was one of seven people to drown that day. Another was Thomas Profitt, a cameraman with Universal Newsreel, who was filming the crowds lining the bank of the river at the time and never even saw the railway pass. His body was discovered still holding his camera.

I looked and looked, but this seemed to be the only re-creation raft trip on record.

I didn't bother mentioning the ill-fated raft trip to anyone in Pembroke during the frantic weeks when the raft was being built. If people were going to forget about red lights / green lights, I was going to forget about a possible, looming disaster.

And sitting on the raft that first day, it seemed even more implausible than when I read about it. The raftsmen in Pennsylvania were still running rapids, not taking into account the rather large obstacles that had been placed in their way since the last time they went down river. We would be lucky, on the other hand, to find a strong current.

Still, on that first day we made our way down the river not feeling entirely secure about the endeavour. Any serious gust of wind knocked the sweep boat around like a fisherman's float. Any wake tossed it around even more. If we were to become untethered we would have to man the sweeps quickly. And the river here was wide with white caps surrounding us as the wind and rain picked up.

We approached Norway Bay and started to see small cottages and houses again. The land started to flatten and clear. A road could be seen on the shore closest to us and we stayed huddled under the tarp eating moose burgers and watching the scenery pass.

Then off to the right, we all saw, at the same time, the oddest sight. It was Sand Point, a jutting point of land that has a lighthouse on it, one of only a handful of lighthouses along the Ottawa River. It is a popular place for amateur photographers to take pictures, and there is a parking lot surrounding it. And there, up ahead, through the rain and the gloom, we saw headlights ringing the lighthouse. As we approached, the headlights started going on and off. Then, through the storm, we heard horns blaring.

Those cars were there for us. I stared in amazement as we rolled by the lighthouse. People had actually come out in the storm, people who likely had never heard about the raft until that morning, had driven down to Sand Point just to see us pass by. We waved as the headlights flashed on and off again. I was left wondering what in the world was going on.

I can't say it was a pleasant first day on the raft, although we tried to make the best of it. From a few light raindrops as we left the mouth of the Bonnechere, a full-blown rain storm descended on us. As the wind picked up, waves started to crash over the bow of the raft. Before long it didn't just feel like surfing on the bow, it felt like surfing anywhere you stood on the raft.

The tarp was good for shelter, granted, but with Tom Stephenson refusing to use it, most of us were finally shamed into going outside. We stood in a line along the bow of the raft, laughing at the wind and rain, drinking warm coffee and eating moose burgers as the waves sloshed around our feet.

After we passed Sand Point, we had only a few kilometres more to travel until we reached the wharf at Braeside, where we would pull in for the night. The river started to narrow. Islands reappeared, and we made our way down the furthest channel on the Ontario side. Before long, we rounded a final point of land, and directly in front of us was the wharf. One by one, people came from under the tarp to stare at what was there waiting for us.

Parked next to the dock were two television trucks from Ottawa, their bright, primary-colour logos clearly visible. You could see reporters with microphones standing next to the trucks, and men with cameras perched on their shoulders standing closer to the water. The parking lot behind them looked full, and there was a bunch of people standing around the television trucks. A woman wearing an apron stood next to a crock pot set up on a fold-out card table.

What in the world?

As we came ashore the cameramen trained their cameras at us, and a loud cheer came up from the crowd. The woman in the apron started talking, and she turned out to be Mary Campbell the mayor of McNab-Braeside who quickly read a proclamation thanking the Shaws for building the "first square-timber crib in 100 years," and how happy everyone in the township was to see one on the river again. Then she served baked beans.

I walked ashore looking for a washroom (one too many moose burgers), and as I made my way through the cluster of people, I shook my head in amazement. Something was happening here. But what?

Late in the afternoon, after a full day on the river and plenty of baked beans in Braeside, the raft started to come apart. It needed to be trucked past Arnprior around the Chats Falls dam at Fitzroy Harbour in order for the journey to continue. The pins were removed, and the timbers hoisted onto a waiting truck. There was talk of trying to get the reassembly to under two hours. Bets were made.

The television crews interviewed Tom and Dana while they worked. You could tell by the way the young reporters' eyes lit up that they couldn't get enough of Stephenson. He hugged the young woman from A-Channel television. Tilted his Jethro hat to a jaunty angle, then talked about carving white pine into oars, or what a seventy-something-year-old man needs in the way of proper hand tools to build a decent raft these days.

I sat there and listened. Wanted to throw something at him from time to time, just to throw off his patter, but sat there and enjoyed it nonetheless. Something about rafts. No doubt about it.

Isle de Montréal, July 31, 1806

It had been a slow journey down river. Although the raft was built with sweeps and rudder oars, and each of the men had poles as well, it was nonetheless a daunting task to steer the vessel. They were forced down channels they did not want to travel. The rafts even crashed into each other from time to time.

They did not know the river, either. They ran aground many times, had to take apart the raft more than once and then put it back together down river. The process had been long, tedious; as dangerous as the rapids were, on some days they were almost to be welcomed.

Not that you did that at the time. The Lachine Rapids, for one, had been everything Tiberius had heard about, everything a frightened sailor had nightmares about. It was the drop in the river that made it so dangerous, a rapid descent that created eddies and miniature waterfalls. It was no wonder Jacques Cartier went just as far as the eastern edge of the Lachine Rapids, and there on his maps, declared it the edge of the known world.

But now, for all that work, it appeared the trip might end in failure. The days lost when they ran aground, or were forced to break apart the raft and rebuild it down-river—they had finally added up. They were running late. His father had a contract to sell the oak staves they carried, but he would miss the deadline if they were not in Québec soon.

So it came to pass that Tiberius stood with his father at the head of a new set of rapids, stretching this time as far as their eyes could see. So much water rushing over rocks, it filled the air with a mist that looked like fog. A stretch of river where the sun never shone. Where the din of water was as loud as thunder.

The North Channel around Isle de Montréal. In the past month Tiberius had seen the Lachine Rapids, the Long Sault Rapids, the Chute de Blondeau, but if there was a more dangerous looking stretch of river in this

world than the one that lay in front of them now, the young man could not imagine where it might be found.

"They say you cannot go this way," said his father, and Tiberius nodded. He had heard the stories for weeks now.

How no boats went down the North Channel. How everyone went to the south. If you were to try the rapids to the north, it would be a suicide journey. And a short one at that.

But the south channel was several days hard poling away. They were already low on provisions. His father had begun to worry, not so much about honouring his contract, but being so late he would miss the merchant ships heading back to England before the winter storms arrived on the North Atlantic.

"What do you think?" asked Tiberius.

"I think," said Philemon Wright, after a long pause, after walking back and forth, looking at the rapids, "that I will not believe it impossible until I have tried it."

They were going down the North Channel.

CHAPTER TWELVE

Quyon, Québec, June 24, 2008

Braeside is several kilometres upriver from Fitzroy Harbour where the Chats Falls Dam was built in 1939. This is the first obstacle we encounter on our trip and, after the raft is broken apart and put in the truck, it is taken to Quyon, Québec on the other side of the dam. Today, St. Jean Baptiste Day, we start putting it back together.

The torrential rains of the day before have passed, but the morning breaks grey and sullen, nonetheless. The sun is hidden behind a low-lying bank of clouds. Cormorants twirl over the water, dipping occasionally to snatch a fish, their black wings flashing. A ferry to Fitzroy Harbour crosses the river next to us every five minutes, the engine being the only sound we hear.

The timbers are placed in the water, the outside crib and traverses first, then they're pinned. The middle beams are placed in one by one, the final piece being stomped into place. It's a perfect fit. All that carving in the pole yard is paying off.

The king's logs are placed on top, and in just over two hours we are bringing the gear back on the raft. Bets are paid off. Those who thought the raft could be put back together in less than two hours insist it will happen in Ottawa. New bets are made.

We will stay in Quyon for the day, head out the following morning. Dana fires up his barbecue. I sit on the raft and read a book.

It is a small book about Philemon Wright, the gentleman farmer from Massachusetts who came north into "howling wilderness" to try and establish an agrarian Eden, only to realize he was on land that would never sustain much more than white pine. As I read, it occurs to me that this country tests you, tests you again, then one last time for good luck.

If you look at our history, how can you deny it? Champlain came with plans of finding a route to China, only to have his journeys cut short by intrigues at the Palace and the latter machinations of Cardinal Richelieu. Sadly, he ended his days as a stodgy administrator of New France, looking at his maps of a "sweet-water sea" and wondering if he would ever return to it.

Two hundred years later, John By built one of the engineering marvels of the nineteenth-century, a 202-kilometre canal system through his own howling wilderness. He built the largest dam in the British Empire at Jones Falls, a structure that looked so out of place, you might as well have stumbled upon an Inca temple in the middle of the jungle.

And for this marvel, he was censured back in England for going over budget. He spent his last years ill, puttering in the gardens of his home near Sussex, dreaming of getting back to the land he had purchased on Parliament Hill, until he finally expired from malaria.

Or what of the 100-acre-land-grant settlers who arrived in Upper Canada in the latter half of the nineteenth century, all those intrepid souls from Scotland, Ireland and England, who disembarked from trains with scraps of paper in their hands and children in tow, only to hike miles into the bush to claim land that would never sustain them, although it would give that illusion for generations?

Tested. Then again. And one more time for good luck.

And of course there was Philemon Wright who came north with dreams of building an agricultural paradise by the shores of one of the wildest rivers in the world. Within six years the dream was in trouble. At that point—like Champlain, By and every settler who came after them—Wright could have packed up and left, could have pulled a Jacques Cartier—who called this place the "land God gave to Cain"—and high-tailed it back to wherever he came from.

But he stayed.

Québec City, August 15, 1806

They had gone down the North Channel of Isle de Montréal with little more than courage and prayers. Any attempt to steer the raft would have been vainglorious. The men tethered themselves to the beams, closed their eyes and slipped down river simply hoping to see another morning.

The sun refracting off the water was dazzling that morning, almost blinding, as the rafts dipped quickly,

water and light wrapping around them like a tourniquet, no bearings or reference points whatsoever, the world turned elemental, voices shrieking in excitement somewhere above the torrent.

And then, just as quickly as it started, it was over. The rafts drifted by the far eastern shore of Isle de Montréal, and the river turned to a calm bath. They could scarely believe they had succeeded. That the last stretch of rapids on their journey lay behind them.

"We have done it, Father," said Tiberius, and Philemon nodded. It would be smooth travelling from here to Québec. No more rapids. With luck, no more shoals. This part of the river was well travelled, had seen timber rafts for years, and with luck, or providence, the worst was behind them.

The Lord rewards the bold. Be either hot or cold, for He spews out the lukewarm. Those may well have been Wright's thoughts at that moment, and he would have been justified in having such thoughts.

So what are we to make of fate? After all that work, all that daring, there should have been a reward around the next bend in the river. If life were a mathematical equation, that is how it would have computed.

But Philemon Wright's fear—of missing his contract, of arriving too late for the merchant ships—was about to come true. When father and son arrived in Québec on August 15 with their nineteen rafts, they found no one wanted to buy their wood.

And so they waited.

It is mid-morning in Quyon, I have put aside my book as people have started to arrive to see the raft. One of the first people on the wharf is Mario Coté, a seasonal fruit picker who lives just outside the village.

"I saw it on television last night," he says. "You guys looked pretty wet."

"Little bit of rain," says Stephenson. "Won't make you melt."

I needed to hear that. Really needed to hear that one more time. But I say nothing.

Coté looks to be in his late forties, with long hair and a ready laugh. He has a fold-out camper's saw that he keeps attached to a belt loop on his pants. He asks Stephenson if he has used ironwood for the pins, and with a bit of dawning respect, he says Yes, that's exactly right. Coté says he used to work in CIP (Canadian International Paper) lumber camps when he was a younger man. Then, when there is a discussion about sawing a couple of inches of one off the flagpoles so it will line up better, Coté jumps on the raft, takes off his camper's saw and sets to work. Don't know how many times a day he gets to use that thing, but he's happy right now.

A trickle of other people start to arrive. There is an elderly couple from Dunrobin who have taken the ferry across. Two men in ball caps arrive in a pick-up truck, get out, and as they walk towards the wharf, they push the caps back on their heads, scratch their temples, and let out a low whistle. Haven't seen that since the pole yard.

By noon, a couple of reporters have arrived, the mayor of Pontiac, and a man from the local Kiwanis branch asking if we need coffee. No OPP-Sureté-Transport-Canada-National-Capital-Commission-or-Fisheries-and-Oceans-people though, so we're doing well.

I sit on the raft and watch people arrive, astounded again by how much interest there is in this ungainly craft with its three jerry-rigged flags and its twenty-seven square timbers. I know a love

of history might bring people out—but it doesn't seem like enough of an explanation.

I watch Coté talk to some of the arriving visitors. After sawing a new flagpole, he has been invited to stay for lunch, and some people are mistaking him for one of the crew. No one corrects anyone. Hell, if he wants to jump on tomorrow, he can. Later in the day, another visitor will say he remembers Coté from working in the lumber camps years ago. Coté was a musician who played some strange trumpet, and every evening you would hear him practising in the bush. Wasn't half bad, as the visitor recalls. For a strange trumpet, that is.

I have an unexpected meeting in Ottawa that afternoon and before leaving Quyon, I pull Coté aside. By now there is a steady stream of people arriving on the wharf, and Stephenson is inviting everyone aboard. You can barely move around on the raft, there are so many people. Coté jumps off and walks with me to my car.

"I hear you might come back tomorrow."

"I might," he says. "I'd like to see you guys off. See this thing moving around in the water."

"It's going to be pretty early," I say. "Are you that interested?"

"Oh, man, I wouldn't miss it."

"Why?"

"What do you mean?"

"What do you like about it so much?"

"The raft?"

"Yes."

Coté doesn't answer right away. He looks back at the raft, and as if he can't help himself, he lets out a low whistle.

"I don't know," he finally says. "It is just so simple."

He doesn't say anything for a minute, and then just before I drive away, as if feeling something more needs to be said, he says:

"It's beautifully simple. You know what I mean?"

I drive back to Ottawa in a mad panic. The crowds that came out to see the raft have distracted me, I've lost track of time, and now I have to get downtown for a three o'clock appointment.

The appointment was made yesterday when I got back to my office and found a message waiting. The Right Honourable Herb Gray wanted to meet me. I phone the number that has been left, and in a few minutes I'm talking to the former Deputy Prime Minister of Canada.

"I have something in my office. I think you'll find it interesting," he says. "After reading your article in the *Sun*, I'm pretty sure you'll find it interesting."

I am an Ottawa boy, and so I make an appointment with Herb Gray. At his convenience. Whatever time suits Herb Gray the best, which turns out to be the following day at 3 p.m..

What am I supposed to say? Herb Gray represented the riding of Essex West (later Windsor West) for forty years. He is the longest, continuous-serving MP in the history of Canada.

He makes a point of emphasizing the word "continuous" when you talk to him, doesn't want to be seen as taking on airs, as competing with Wilfred Laurier (forty-five years in the House, booted out once) or William Lyon McKenzie King (Canada's longest-serving Prime Minister, booted out twice.)

But as for continuous electoral victory? Yes, that's Herb Gray. The man who couldn't lose so much as a beauty contest in Windsor. Who, in the years he was Deputy Prime Minister under Chrétien, used to out-poll his boss in popularity enough times for it to be uncomfortable.

Nothing flashy. Nothing over-the-top. That was never the Herb Gray style. And that stolid earnestness made him a media celebrity in the nineties. A Drew Carey look-a-like in a land of slicked-back hucksters.

So we make an appointment at his offices — "It's the International Joint Commission," he says — and I say Yes, I'll be there, hanging up the phone, and having not a clue what the International Joint Commission is.

I figure it out that night. The International Joint Commission is a bilateral organization that deals with Canada-United States transboundary issues on water and air rights. Gray was appointed the Canadian Chair in 2002 when he retired from the House.

Anyway, I roar my way down the 417, cut through early commuter traffic downtown and find myself being buzzed into the offices of the International Joint Commission, at five minutes past three. There is a large reception area tucked behind glass, a busy photocopier behind it, people walking to and fro with papers in their hands.

I sit in a reception area and wait. Before long I see Herb Gray making his way around a corner. He doesn't look that different from the last time I saw him on television. The shoulders are perhaps more stooped, the walk a little slower. But the hand-grip is firm, and he wastes no time in leading me to his office.

We chat as we stroll down the expansive corridors of the International Joint Commission. The lousy weather so far this summer. The rumours of a fall election. I ask if he misses the House.

"Oh sure," he says, "but it was time to go. And I'm busy enough."

Which leads us to talking about the International Joint Commission, from there, to the Canadian-American border, and before long we are talking about water.

"Forty per cent of our border with the United States is water," says Gray. "Not many people realize that, but look at a map and you can see it clear enough."

He points to a painting on his office wall. The waterfront in Windsor.

"That was my riding. The lake was part of my childhood. We had a cottage as well. We went there every summer, and it was always something I looked forward to. I've always loved water."

We talk about the waterfront in Windsor ("It's a miracle what they've done; I go back there today, and I don't recognize it as the same place."), then about cottages, and as I'm sitting there wondering exactly why I'm here, Gray starts to talk about something called the Winkworth Collection.

"The Archives came to me and asked for my help in acquiring the Winkworth Collection," he says. "As soon as I saw what was there, I enthusiastically said Yes."

I sit there not understanding a word, although in a few minutes I piece together the story. Peter

Herb Gray poses for a photo at his office with a reproduction of a painting by Frances Ann Hopkins showing a large timber raft. (Sunmedia photo by Errol McGihon)

Winkworth was a wealthy collector of Canadian art who by 2002 had assembled perhaps the largest private collection in the world.

Gray helped raise the $5.3 million the National Archives eventually spent to acquire the collection—3,200 prints, 700 watecolours and drawings, more than 100 maps.

"It's why I called you," he says, "I was thinking of the Winkworth Collection when I was reading your story about the timber raft in Pembroke. Come, I want to show you something."

We leave his office and make our way down a hallway. The former Leader of the Opposition, former Deputy Prime Minister, ten-time cabinet minister, longest continuous-serving MP in the history of Canada, is speaking as we go:

"The Archives gave me a copy of a Winkworth painting when I retired from the House. No one even knew the painting existed until 1996."

We are slowing down; Herb Gray is still talking: "I have always loved the painting. I put in on display here so people can see it."

And we are finally there. In front of me is a painting by Frances Anne Hopkins dated 1870, and I can scarcely believe my eyes. I'm staring at a timber raft. A big one.

"You're right," says Herb Gray, "when you tell people how important rafts were to this country. We need to remember our history."

I nod and agree as we stare at the painting of a timber raft, both of us, I suspect, wishing we were on it.

Quyon, Québec, June 25, 2008

It is early morning, the ferry has just started running, and we are back on the wharf at Quyon, getting ready for another day on the river. Although I don't believe there is even a restaurant open in Quyon yet, people have come to see us depart.

Mario Coté has come back, this time with a voyageur's scarf attached to a different belt loop. He has also found a hat almost identical to Stephenson's, looks every bit a raftsmen, although he is not accompanying us on the river. He sadly explains he would have no way back home once we moored for the night. Maybe he'll come see us in Ottawa.

Eddie McCann the mayor of Pontiac has come back as well, passing out lapel pins and offering coffee. He says if he had more warning that we were coming, he would have arranged a night at Gavin's, a legendary drinking establishment in the village.

"Call next time," he says. "We'll take you lads out for the night."

As I load the coolers onto the raft, I see another group of people arriving. Two men and a woman. One of the men is older, and when they reach the raft, he is introduced as Carl Trudeau, the woman's uncle as well as an Ottawa Valley fiddle player who used to perform with the Quyon Old Time Fiddlers.

"Uncle Carl really wanted to see this thing," the woman says as Trudeau stands beside her nodding his head.

"I read a story about it," he says, looking down at the raft. "It looks even bigger when you see it up close."

That night I am hosting a live radio show from the raft, and I am curious about Trudeau's fiddle playing. He says he learned to play from his father who learned the fiddle while working in the bush camps. He never bothered with that job—other than a few winters when he was a much younger man—but Trudeau learned to love the fiddle all the same.

"Never had any complaints," he says, when I flat-out ask how good he is. I wait for him to say something more, then feel like a fool as we stand on the wharf not speaking to each other.

We are heading downriver to Ottawa, I quickly say. We're doing a live radio show that night in Dunrobin. Why doesn't Trudeau bring his fiddle and join us? It would be nice to have some music.

"Would need a guitar player," he suggests.

"Bring a guitar player."

"Might need a step-dancer."

"Bring a step dancer."

Trudeau nods. You can see he's thinking about it, even though he's not ready to commit. Good Upper Ottawa Valley way of looking at things.

"Give me a call this afternoon," he finally says.

Not at all sure I'll have cellphone reception where I'm going, but I say okay, then give him the address where we'll be spending the night, say again it would be fun to have some fiddle music on the raft. And we certainly have room for dancers and back-up musicians. Knock yourself out.

We keep loading the raft, Coté helping out, as the first raindrops of the day start to fall. Dana Shaw comes over and says it's time to go. As we step onto the raft, people on the wharf clap and cheer, and we row our way to the middle channel, making slow progress until we round a point and find a bit of current.

The current at the mouth of the Quyon River doesn't last. Before long, we tether the raft to the sweep boat and continue our way towards Ottawa.

We travel down Woolsey Narrows, the thinnest part of the river we will pass through. The pine towers over us on both sides of the river, the water momentarily cast in shadows. The contrast from Castleford where we started the trip, where the river was so wide, and where there were so many islands you might have thought you were catching the ferry to Victoria—it couldn't be more dramatic.

Then the river widens again, and off to our right is the yawning entrance to Buckham's Bay, narrowing into a spit of water far down the opening of the bay. It looks intriguing, what we might find if we go that way, but we have to stay on the main channel.

As we approach Constance Bay, the river becomes wide enough again to be a lake. We see people swimming in the distance. Hear laughter and children's screams. Far downriver I see a speck of land in the middle of the channel—Twelve Mile Island it says on my map—and we make our way towards it.

As I sit on the raft I imagine what it must have been like for the first raftsmen to travel this river. Fatalities were common. People disappearing downriver and never returning were common. Before long, as I become accustomed again to the gentle roll of wood on water, as I watch Dana lift the lid of a barbecue, and listen to Tom tell stories to Gouge and his friend, the RCMP officer from British Columbia, I find myself lost in thought.

For those people who love rafts, for many of them, it probably has something to do with Mark Twain. And Huckleberry Finn. And one of the greatest stories ever written. I read it last week, just to remind myself.

Twain published *The Adventures of Huckleberry Finn* in 1884, his twentieth published book, and the one that cemented his reputation as one of the finest English writers of his generation. Although banned over the years by various school boards (because of its open questioning of authority, or alternately, his word choices, specifically his descriptions of the negro slave Ol' Jim by using the vernacular of the day), the story of the self-reliant boy, the escaped slave and the raft they took out onto the Mississippi River has attained almost fable-like status.

It is no coincidence that Twain picked a raft as the locale for his story. His love of the vessel is almost palpable in the book: "Sometimes we'd have the whole river all to ourselves for the longest time. Yonder was the banks and the islands across the water; and maybe a spark — which was a candle in a cabin window; and sometimes on the water you could see a spark or two — on a raft or a scow, you know; and maybe you could hear a fiddle or a song coming over from one of them crafts. It's lovely to live on a raft."

How much did Twain love rafts, steamboats and rivers? Consider this: his very penname comes from the Mississippi River. Born Samuel Langhorne Clemens in 1835, he started using the name Mark Twain after 1863. The name is what Mississippi river-men used to holler when they wanted to indicate a safe depth for a boat to float over, twain being an archaic term for two, and two feet being the desired depth.

"Mark twain" the young writer would hear, whenever he was on a steamboat and a safe sounding had been taken. The words stayed with him, as surely as the river, as his love of rafts did, for the rest of his life.

"What you want on a raft above all things," Twain wrote in *The Adventures of Huckleberry Finn*, "is for everybody to be satisfied and feel right and kind toward the others."

CHAPTER THIRTEEN

June 25, 2008

I'm sitting on a back beam with Tom Stephenson, everyone else on the raft gathered around the barbecue. The rain has been steady today, but without the force of yesterday. There are even patches of blue sky ahead, and talk of all this bad weather clearing before the radio show.

When Tom Stephenson graduated from the University of New Brunswick, he returned home, but soon left Perth to begin working in lumber camps around Pembroke and Fort Coulonge on the Québec side. He spent a lot of years working for the Gillies Brothers, the old competitors of the Shaws. He must have been a star hire—he was tall and strong, didn't mind the odd scrap on a Saturday night, easily commanded the respect of the men under him, and before long, he was a camp foreman.

Not many of the men in the bush ever knew he had a university degree. They knew he had some schooling all right because he used to help some of the older men with their writing and reading, even gave ad hoc lessons Sunday morning for anyone who wanted them. Some of the men in the camps could not even sign their names. So everyone knew Tom had some schooling, but they were nonetheless shocked when he quit the camps to become a college teacher. No one thought he knew that much.

"I thought it would be putting on airs if I talked about it," Stephenson says, sitting on the back of the raft. "People in the camps judged people by what they did, not by the letters after their names.

"That was one of the things I always liked about the camps. That and the food. That you could just show up, do your job, and you'd be accepted."

So he never talked about his four years at university. Just went about his work. Even when he met his wife Judy she had no idea she was dating a "college man." He lived on the same street as her, Judy's sister delived his newspaper, and she remembers thinking at the time: "He's cute, but what a hick."

Tom was smitten, though, and he started to pursue the young English teacher. He would offer to take her skiing, or out for canoe trips in Algonquin Park. He knew a few spots, he would say. He wore her down, and before long they were dating. He was athletic, and she liked that. Most anywhere they went, he was the tallest in the room. The most animated. The person having the most fun. His enthusiasm was infectious, and Judy started to look forward to their time together. In 1966 he proposed, and the following year they married.

It couldn't have been easy those first years with Tom in the bush for months at a time, or after that,

Tom Stephenson is silhouetted against the clouds and blue sky after hanging his socks to dry as the square-timber crib makes its way along the Ottawa River on Wednesday, June 25, 2008. (Darren Calabrese / Courtesty of the Ottawa Sun)

when he started working for the Ottawa River Forest Protection Association, putting out forest fires up and down the Valley, or anywhere else in the country that needed urgent help. He was always restless, although he never wanted to be any place but the bush. Judy grew up with many people like that—you find them all over Pembroke—so it didn't surprise her greatly. And it wasn't like Tom was so full of his crazy ideas that he didn't see the wisdom in taking that college job for a few years so he could help with the children. If she had asked him, he would probably still be teaching.

But she wouldn't do that to him. She had never known a harder worker than Tom, so she never worried about paying the bills. And if you weren't worried about money, well, that was a form of freedom wasn't it? You should use it.

So Tom went back to the bush. Went back to fighting forest fires. Their daughter Jennifer grew to have her father's love of the outdoors and worked as a cook in the camps before trying her own hand at fighting forest fires. She lived in Wasaga Beach now, was married, was starting to raise a family. Their son John had the same love of the bush, and he hadn't gone far; worked as a chiropractor in Deep River, and they saw him every other day. They had three grandchildren altogether, two with John, one with Jennifer, and Tom doted on the children.

Tom had been a good husband even though they spent less time together than if she had married a combat soldier. Men in the bush. Once again, Judy wasn't surprised. It's like the raft. Once he started, she barely saw him. He said he'd phone when he reached Ottawa. And she knew he would.

Meanwhile, Stephenson sits on the back beam, laughing and refusing to tell me how old he is. The other day, Judy told me he was seventy-one.

Dunrobin, Ontario, June 25, 2008

Shortly before 4 p.m. the rain stops and the clouds start to clear. The sun never fully arrives, but the last leg of the day's trip is spent outside the tarp, not getting wet, and that seems novel enough.

We keep chugging our way to Dunrobin where we will moor for the night at the beachfront home of Mike McGann, a friend of the Shaws. We pass Twelve-Mile Island. See the Gatineau Hills starting to form on the northern shore. Boats honk their horns as they pass us. A few stop, the drivers come onboard the raft.

"I want one of these," a young man driving a bullet boat says when he steps onto the raft. "Talk about a party boat. You could put an entire band on this thing."

Shortly after five, we man the sweeps and make our way toward shore, rowing past million-dollar homes on our wonky raft, as we look for Mike McGann's backyard. When we see a crowd of people and the van from the radio station, we know we've arrived.

There in the crowd, I see Carl Trudeau standing with a fiddle case, next to him a man carrying a guitar case. The guitar player's name turns out to be Dennis Bennett and, although Trudeau hasn't found a step dancer, he's in luck because we've brought one with us. Paulette Gauthier, who has been riding in the sweep boat with her friend Peter Haughton, has been with us all day. She teaches step dancing. Even plays a little fiddle.

In the back of the crowd, I see Ward Stewart from CFRA who has already started to lay down cable and set up microphones. It's a crazy idea, trying to do a live remote from the raft, no one is even sure if it is going to work, let alone what might happen if we get hit with a wake, and water rushes over the cables and equipment. That being said, this whole idea started with people wondering if it would work, so it doesn't even seem like cause for stress anymore.

I help Ward set up a table and chairs on the raft. Check for cell service. Don't have any. Borrow a phone from McGann and phone the studio. We're going to have set up a land line. Here's hoping that wake story never happens.

As we work, more people start to arrive by the dock. John Shaw has driven down. Some friends of Trudeau come walking around the corner. Before long, there is a festive mood on the raft. Someone opens a beer. Someone else asks if you can have a beer before a radio show. Another person yells out it's radio, not television, and someone opens another beer.

At 8:05 p.m., with fingers crossed, we flick a switch and go live from Mike McGann's backyard.

"Good evening and welcome to "Unscripted." I'm Ron Corbett, and welcome to a special show

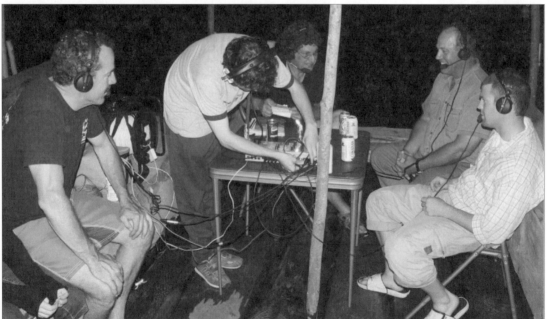

Top: *Tom Stephenson and Nicole Shaw giving a radio interview on the raft.* Bottom: *Mike McGann, Ward Stewart, Ron Corbett, Jean Gougeon and Paddy Irving sitting on the raft during a live radio show. (Dana Shaw)*

tonight, live from a square-timber raft on the Ottawa River."

And just then a boat passes close to shore, no doubt attracted by the lights and commotion, the wake sloshing up on the raft and swirling around my feet like an incoming tide.

I believe my next words were: "Christ, we already have water."

The five-second delay worked without a hitch.

As it turned out, that was the only boat to pass close enough to throw up a wake. The rest of the show went off without incident. Tom told his story of the banker he used to carry on his back so the man could go fishing, finishing the story by saying that was about the only good reason to ever carry a man around on your back.

Trudeau, Bennett and Gauthier played, danced, and there was loud applause after every performance. People started phoning the studio in Ottawa to make requests, which were then forwarded out to us. Our impromptu trio played "The Log Driver's Waltz," Irish reels, and kept taking requests.

Dave Lemkay talked about the heyday of rafting on the Ottawa River, the days of the Shiners and Big Joe Muffraw when raftsmen fought tooth and nail for the right of being first to get down the timber slides. If you were too late getting to Québec, you might not find a buyer. If you were stuck without a buyer, you might as well sink the logs right in the Québec Harbour. That's how useful they were.

Once, just before a commercial break, someone I didn't even know who was fishing from the raft, caught a catfish. He pulled it from the water in a grand, sweeping arc, the fish landing on the table in front of me. I kept saying, "We'll be right back to "Unscripted" after these quick messages from CFRA," as I stared down at the fish's whiskers and watched its tail flop around a headset.

Mark Twain, you should have been here. You would have loved it.

Québec City, October 21, 1806

Philemon took the book from the breast pocket of his coat and placed it on a table. He had promised himself he would write an entry every day to chronicle his trip down the river.

He had kept meticulous records at the start. How many trees he was bringing with him. How much salted pork and flour. The names of the men he travelled with, or whom he hired along the way.

The shoals, rapids and eddies they encountered, he made a note of those. How many rafts had to be taken apart, to be assembled later downriver. More notes.

When the trip took longer than expected and the party ran out of food, he made notes on what he purchased, how much it cost, and by the end of the trip, the debts he had incurred.

That last part took up too many pages in the book. He looked at it in dismay. This venture had the potential to be a debacle, now that he was lodged in Québec City still taking on debt with no buyers anywhere for his trees.

They had arrived too late. That's what he and his son had been told. The loggers in Lower Canada had already stocked the merchant ships returning to England.

He could stay until the snow fell, but would have to leave for home after that. If he waited any longer, he would be stuck until the rivers froze. With snow in the bush and no water frozen, there would simply be no way home.

He opened the pages of his book and found his last entry. It was the same entry he had penned every day for two weeks. He could not think of anything else to say, and so he picked up his quill and wrote the words one more time.

"Waiting to sell."

Jean Gougeon and Tom Stephenson sit on the back of the raft on the last morning of the trip. (Dana Shaw)

CHAPTER FOURTEEN

Dunrobin, Ontario, June 27, 2008

For the first time during our trip, the day starts without a cloud in the sky. You can see the sun rising downriver, a soft band of yellow light that gradually spreads and then takes shape. The shadows retreat. A slow mist begins to circle on top of the water.

We sit on the raft looking at the sunrise. By the end of the day we will be in Ottawa, and there is a certain melancholy to that.

Tom adjusts a flag at the front of the boat ("The Canadian flag should be a foot taller I think, at least.") while Dana inspects the contents of a cooler. Gouge lies on his back, a hat pulled over his eyes. Dave Lemkay fiddles with the motor of the sweep boat. Scott McLellan, a friend of Tom's, who stands six-foot-eight, weighs more than 300 pounds, and could easily be a double for Big Joe Muffraw, has also joined us for the last day.

No one seems in a hurry. Truth be told, we had been leaving later and later every morning. The river had taken away any sense of urgency—the rush to assemble the raft, the concern about getting it done on time—all that seemed a long time ago.

Drifting down river, even when being pulled by the sweep boat, it seems like you have all the time in the world. Deadlines can wait. Destinations will present themselves. Nothing seems of such import that it needs to be done immediately, and the hours spin out as slow as the maple syrup Tom taps in the Lanark Highlands.

By 10 a.m. we are finally ready to start the day. Lemkay pulls the raft to the middle channel where we start to move downriver.

It is nice to finally have a day without rain, and we take advantage of it by sitting on the back of the raft, letting our feet dangle in the river. The shoreline now is a steady flow of large homes, sometimes larger docks. In the distance, the Nortel campus can be seen.

We will soon be back in Ottawa, my hometown, cutting through the city on a raft.

I look on a map and see Mohr Island is ahead. I've written about Mohr Island before, a "very pleasant island" according to Samuel de Champlain who stayed there on both his voyages up the Ottawa River. It would have been a natural overnight stop, leaving a day's journey before reaching the shock-and-awe of the Chaudière Falls. There, it would have been the custom of his Algonquin guides to spend the night and make offerings to the gods before continuing their journey.

Given the history of the Ottawa River, it is a mystery to me why it has never been given heritage status, like many rivers around it. Fred Blackstein, who helped with the raft project from the first meeting, heads up a committee that is trying to get such a designation.

It was the Ottawa River that Jacques Cartier saw when he walked to the top of Mount Royal in 1535, a mighty river stretching to the western horizon, one he could only dream about because there was no way of reaching it, no way of getting past the Lachine Rapids.

Not until 1608 did a European finally reach the shores of the Ottawa River, and until that time, it became a mythical place, a possible route to the Orient that could be seen, but never reached. Almost a vision. It was given the name, la Grande Rivière du Nord.

The first European to finally reach its shores was a young Parisian street urchin by the name of Etiènne Brulé. It was Brulé who was there a couple of years before the historic firsts normally given to Champlain: the first European to travel up the Ottawa River, to see the Great Lakes, to see the present-day site of Toronto.

Brulé fell in love with this New World, and became one of the best interpreters Champlain had under his command, although he ultimately betrayed Champlain to the English, and for that crime was murdered by the Hurons. His bones were never found. You find human remains on the Ottawa River from time to time. At Sand Point. On Mohr Island.

Tom Stephenson is silhouetted against rain clouds while steering the timber crib along the Ottawa River before landing just upstream from Gillies Old Mill in Braeside on Monday, June 23, 2008. (Darren Calabrese / Courtesty of the Ottawa Sun)

We chug along, watching the city grow around us. As hard as it is to believe right away, it actually seems to be getting warmer.

You tell stories on a raft. If you're good at it, the stories don't drag, become preachy, or meander into predictable "you-know-what-I-think" territory, which is the conversational equivalent of being in a car when it drives off a cliff.

Tom Stephenson is pretty good at telling stories.

"There was a whore house near Fort Coulonge that operated for years," he will say. "There was talk of closing it down once, but even the ladies in town didn't think much of the idea. They had the best breakfast in town."

Or, another time: "I got in a fight once and landed a lucky punch. Dropped the fellow like a cord of wood. Don't know how it happened, but no one messed with me after that. I told everyone, of course, that I knew exactly what I was doing. Watch out for me."

After a few days on the raft, everyone started telling stories. Dana told a story about a hunter who thought he knew everything, and so one day at camp, they pulled a prank on the man, leading him to a spot in the forest where they had hidden one of the trophy plaques from the hunt cabin. When the man saw the buck's head poking out from the trees, he got so excited he almost dropped his rifle.

He fell to his knees. Aimed and fired. Fired again. When the head disappeared in the bushes (knocked out of the tree) he figured the animal had taken off. He started to chase it. Yelled for Johnny Shaw to follow him. But Johnny couldn't. He was too busy rolling on the ground laughing.

When the man finally figured out what was happening, he swore and blustered that he had been playing along, knew what was happening the whole time—although when they put the plaque back on the wall of the cabin with a new brass tag underneath explaining the trophy buck had been shot in 1995, then "reshot" by the man in 2003, well, let's just say he hasn't been seen around the hunt cabin much.

A lot of stories. Here's one more. It's probably time it got told. The story of how we found our final destination.

Three weeks earlier, when the raft was starting to take shape in the pole yard in Pembroke, we finally got around to arranging the actual trip. The sad, sorry truth was we still didn't know where we were going.

After much discussion, it was decided to try and take the raft to Ottawa. As I was the "Ottawa connection" in this little venture, I was conscripted to make some calls. See if there was a place we could moor for a night.

That was the original plan. Find a place to stay one night, then back on the truck and back to Pembroke. We do that, and we can say we did it. We took a square-timber raft to Ottawa one last time for the world.

I phoned the Hull Marina where a polite young woman told me they didn't provide slippage to timber rafts. Ditto for the next marina. And the one after that. One marina did ask how big the raft was, but once the woman was told, that was a short conversation as well.

I started to think of other possible sites. Racked my brain. The Rockliffe Rowing Club? How about that crazy old boathouse just downriver from it? What about a beach? Could we just drop anchor off Westboro Beach?

I made a few phone calls: "What is it you want to moor at Westboro Beach exactly?" I was quickly getting nowhere. For a city that was founded at a place where three rivers meet, there aren't as many places to park a raft as you might think.

There was one site, however, that I thought of right away. I just didn't call it. Last time I called this place to talk about timber rafts, it didn't go so well. I amused the heck out of the person I was talking to, but that was as far as it went.

But finally, after a week of being laughed at up and down the banks of the Ottawa River, I decided I could take it one more time. So I picked up the phone and dialed the main number for the Canadian Museum of Civilization.

This time, I was put through to a Yasmine Mingay, the head of communications for the museum.

"It's a raft?" she said after I explained what we were doing.

"That's right," I answered. "One of the old square-timber rafts."

"And you need a place to moor it?"

"Somewhere in the city, that's right. You have a wharf I seem to remember."

"We did. It was taken down last year. It might still work, though."

Might still work? Did I just hear that?

"I bet you it would," I said quickly. "You just need a way to anchor it. You don't really need a dock."

"Interesting," she said. "Listen, can you send me some information about this?"

I promised I would. Then hung up the phone and called Dana Shaw.

I repeated the conversation to him word for word.

"Interesting? That's what she said?"

"That's what she said," I answered.

And that's all she said. A little window of opportunity opening. Maybe. And yet when I got off the phone, Dana did the craziest thing: He went and got Clarence Lorbetskie—perhaps the finest truck-driver-cum-loader-operator in Eastern Ontario—and the two men drove to the Canadian Museum of Civilization.

There, Dana and Clarence had a look at the wharf that was no longer a wharf. At the service roads leading into the museum. At the Waterfall Court, the stairs leading to it and the different walking paths by the river. There were a lot of people around, but the two men barely noticed them, so intent were they on the survey job.

After that, they walked through the front doors of the museum, went to the woman selling admission tickets and introduced themselves.

"Is Yasmine around?" Dana then asked. "If she is, can you tell her Dana Shaw and Clarence Lorbetskie are here? Tell her we've had a look, and we think the raft will work just fine here."

Crazy story. We've laughed about if for days. Craziest part of all is it worked.

Not only worked, but Yasmine Mingay fell in love with the raft. Then with the people who built it. She jumped through hoops and over hurdles to make things happen. Got a lot of other people down at the museum doing the same thing, refused to listen to the "maybe-next-year" people who came to her office and asked if she was feeling all right.

Indeed, by the time we set off on our trip a mere two weeks later, arrangements had been made to put the raft on display at the museum once we arrived. At the Waterfall Court no less. For a week. Where people could see it properly. The dates would overlap with Canada Day.

The final bit of good news came while we were actually on the raft that final day. An e-mail arrived on Dana's Blackberry. A press release from the museum. With all of us laughing and hooting, he read it aloud:

"The Canadian Museum of Civilization is paying an unusual tribute to the Ottawa Valley's rough-hewn past by displaying an authentic squared timber crib. The enormous contraption is on display until July 6 in the Waterfall Court, just outside the museum, following its long journey down the Ottawa River.

"Squared timber cribs were once the only way to ship wood from the logging camps of Ontario and Québec, down the Ottawa River and all the way to Québec City, where it could be loaded onto ships sailing for Great Britain. Unlike boats built to carry cargo, the cribs actually were the cargo—huge timbers assembled into floating structures that were lashed to dozens of other cribs to form an enormous raft of wood.

"This year is the 100th anniversary of the last squared timber drive down the Ottawa River, so Pembroke lumber company Herb Shaw and Sons spearheaded an effort to commemorate the event by building a new crib. Vintage tugboats are towing it to several communities this summer to celebrate the Ottawa Valley's colourful past

and to draw attention to the enduring importance of the Canadian lumber industry, and its move towards greater sustainability.

"'The nineteenth-century river drives opened up this part of the country,' says Victor Rabinovitch, President of the Museum of Civilization Corporation. 'Both Ottawa and Gatineau began life as lumber towns. Even the Canadian Museum of Civilization was built on the foundations of nineteenth-century lumber mills.

"'We are proud to partner with Herb Shaw and Sons in bringing their crib to the National Capital, especially during the week when we celebrate Canada Day.'

"The crib, which measures twenty-four feet by thirty-two feet, was built out of twenty-seven white pine timbers donated by Herb Shaw and Sons, the oldest continually run family lumber business in Canada. It took local expert Tom Stephenson and two Algonquin College students about a month to build the structure. Using traditional methods, they pinned the corners and drove hardwood wedges between the timbers to hold them snug.

"Their old-fashioned masterpiece has been assembled on the grounds of the Museum of Civilization, where visitors can come explore a fascinating and turbulent chapter of our nation's history."

When Dana finished reading we broke into laughter. Giddy, vindication laughter.

"Keep that," screamed Tom. "Any cops that stop us now, you just show 'em that!"

CHAPTER FIFTEEN

June 27, 2008

There is no doubt about it anymore. This is a beautiful day.

As Dana fires up the barbecue for our last meal on the raft (it will be a rather pedestrian offering of beef burgers), people take off their shirts and lie in the sun. Gouge jerry-rigs a diving board at the front of the raft by wedging one of the sweeps between the traverses and cannon-balls himself into the river. Tom strips to his underwear and one-ups him. He does a back flip.

I stare at him and shake my head. Can't think of too many other men in their seventies—unless they were professional swimmers at one time—who would be doing that. When he pulls himself out of the river for another go at the diving board, he almost runs to the sweep. Then he back flips again into the river. Amazing.

We swim for an hour or so, a rope soon tied to the back of the raft so people can grab onto it and pull themselves back aboard. After swimming we eat hamburgers. Lie down on the raft and almost fall asleep. I start to think about what it was like for raftsmen who did this for real. When you worked on the rafts, you earned your money when you were going down the chutes, poling you way around some

shoals somewhere, or breaking all the wood apart in Québec.

Many of the days, though, must have been like this. Just drifting downriver. Nothing to do except think of reaching Québec, getting paid, then making your way back home. A lot of restless, tick-scratching boredom.

Which maybe explains the mayhem that followed whenever the raftsmen came to Ottawa.

At the timber slide that Ruggles Wright designed and built at the Chaudière Falls in 1829, there was always a bottle-neck of rafts waiting to go downriver. Sometimes the raftsmen would be stuck for days. And in that time spent drinking at taverns like Mother McGuinty's or the Exchange Hotel, definite opinions were formed on which rafts should be the next to leave.

"Most of the fights in the taverns and along the river were because of the rafts waiting to go downriver," Dave Lemkay told me one afternoon as we sat on the raft. "If you reached Québec too late in the season, you might not be able to sell your wood. Your entire season could depend on how quickly you got down the timber slides."

Tom Stephenson does a back flip into the Ottawa River (Darren Calabrese/ Courtesy of the Ottawa Sun*)*

The sheer simplicity of it—first man down wins, no prizes for being late—led to a brief period of anarchy along the Ottawa River. It began in 1832, with the completion of the Rideau Canal, which threw thousands of Irish labourers out of work. Most tried to get work in the only other industry around—logging—but few were hired. French Canadians had a near stranglehold on the lumber camps and the annual raft trips to Québec.

There was one man, though, who was willing to hire them. Peter Aylen had arrived in Canada under mysterious circumstances, quickly changed his name (he was born Peter Vallely) and in a few years ended up owning a timber limit along the Madawaska River. Aylen was an ambitious businessman, and his business plan was brutally simple: Control the Ottawa River.

To do that, he turned his recently hired Irish workers into a near street gang. The "Shiners," as they came to be called, assaulted other loggers. Set fire to their rafts. Burned down their farms. Stole wood from their timber limits. When in Bytown, they terrorized the village. People were assaulted on the street. They were mugged. Men from competing lumber companies disappeared from the taverns, never to be seen again, the rumour always being that Shiners had dumped their weighted bodies into the river somewhere.

The gang's control over Bytown was so complete that when Aylen was arrested and detained once, Shiners descended on Ottawa Lock Station and burned a steamboat. Their leader promptly had all charges dismissed against him.

It was only after Aylen attempted to murder prominent Uppertown lawyer James Johnston—who had written to the Lieutenant-Governor to complain about the violence in Bytown—that paid constables were finally brought to town and the Shiners brought under control. This was the beginning of the Ottawa Police Service.

As for Aylen, he was never convicted of any crimes. When the police arrived he simply crossed the river, built a fine stone house and helped settle the city of Aylmer. In 1846, he was elected to the municipal council. He was briefly a justice of the peace.

Many historians have said the "Shiners' Riots," as this brief period of lawlessness came to be called, had much to do with ethnic and cultural hatred between the Irish and the French.

But this ignores a far simpler explanation. It might have been all about the wood.

Funny to look back on history and speculate on how many things we may have got wrong. Or missed by just a little, but out of that error grew centuries of misinformation. A silly fantasyland yawning before us that we believe is the truth.

Still, there is no disputing what happened after Philemon Wright sailed his first raft down the Ottawa River, something many other people would never have considered. The following year, he had no trouble selling his pine. Before long, word started to spread of the vast stands of white pine in the Ottawa Valley.

Within a few years, people started to head north. In the beginning they came as small parties. Peter White and his family arrived in 1828 and started felling trees near Morrison Island. Their homestead became the city of Pembroke.

Daniel McLachlin arrived and built a sawmill at the mouth of the Madawaska River (Arnprior). John McDougall arrived and built a gristmill (Renfrew). John Egan settled near his timber limits on the Bonnechere River (Eganville).

After the small parties, people started to come by the thousands. Then the tens of thousands. By 1847 (the year of the Great Potato Famine), 38,781 people had emigrated to Upper Canada from Ireland alone. A great many of those people headed for the lumber camps of the Ottawa Valley.

Which gives us a last fleeting thought on the often random nature of history. For what Wright didn't know when he came in search of land for the perfect agrarian commune was that he had stumbled upon a near perfect band of land for growing pine.

You notice the difference between pine grown in North Bay or Upper New York State and pine grown in the Ottawa Valley. Climatic and geological conditions came together to create a perfect greenhouse in the Ottawa Valley for white and red pine. Ours is better. And has been for two centuries.

An odd little fact. That changed a great many things.

We have an urge to stretch our legs and so we've stopped at Pinhey's Point, one of those quixotic spots you find around Ottawa with a story to tell of someone who came to the New World with grand plans that didn't quite work.

The crew on the square-timber crib begin the final leg of their journey along the Ottawa River early Friday morning, June 27, 2008. Photo by Darren Calabrese (Darren Calabrese / Courtesty of the Ottawa Sun*)*

The ruins of a carbide mine in the Gatineau Hills. An abandoned lumber camp somewhere in the Ottawa Valley. Any old, sagging cabin you see in a field. The region is littered with such spots.

This one is where wealthy English merchant Hamnett Pinhey was given a land grant in 1820. He set out to build an English estate in a style that was grand and old-fashioned even before he started construction. He would work on the building for a quarter century with the intention of bequeathing it to his son, Horace.

After the first phase was completed (a relatively modest cottage to which would be affixed stone additions), Pinhey started farming. Orchards were his intention. And then the cruel reality of the land set in. The soil was shallow and thin. The northwest wind blew across the river daily from the Gatineau Hills.

Even his family abandoned the estate for more welcoming places. By 1971, only Ruth Pinhey, an elderly spinster, lived in the home. When she died, the home was given to the City of Kanata, now the City of Ottawa, which keeps it open as a museum.

We stroll through the gardens, past the stone manor house, look at stunning vistas of the river. This should have worked.

Plain truth about life some days—it just doesn't.

We stay at the Pinhey Museum for about an hour, then go back to the perfect, sheltered bay where we moored the raft. It is a popular anchorage spot for sailors, and there is a boat there now, a sleepy middle-aged couple standing on the deck rubbing their eyes and taking a second look at what tied up beside them while they slept.

We climb back on the raft, give them a wave, row out to where the sweep boat can tie on, then make our way back to the main channel. Before we round the point, I start thinking again about all the work that has gone into this raft, and wonder, not for the first time, why these men bothered.

I mean, let's be honest here, I'm not travelling with a boatload of Modified-American-Plan historians. This has not been a pleasant boat tour of the logging sites along the Ottawa River. A jaunt on a bus somewhere to see the Bonnechere Caves.

While the journey may have started when these men were told the last timber raft to travel down the Ottawa River made the trip 100 years ago—a love of history couldn't explain everything that happened afterwards.

Perhaps it was the challenge, people saying so many times it couldn't be done, that they finally decided to, just to spite them. The sort of thing people have been doing around here since farmers tended cash-crop farms in the Gatineau Hills or the Algonquin Highlands for generations after they should have left.

So maybe it was the challenge. Or the adventure. I think the adventure appealed to all three of them. After all, how many summers do you get a chance to build a square-timber raft and take it down the Ottawa River? One for the books, as they say up the Line.

And yet that still didn't explain the excitement in the three men's eyes whenever they stepped foot on the raft. Or the obvious fondness they had developed for the ungainly vessel ("I still think we need to make the centre flagpole a little higher. What do you think, Tom?")

It also did nothing to explain the crowds that appeared along the river to see us. The trucks flashing their lights and honking their horns off Sand Point in the middle of a rain storm; the peope who came out to Braeside. Quyon. Why did these people end up caring so much about a wonky old raft?

It was a riddle. A pleasant mental exercise to spend the day on as I sat on the back of the raft watching Pinhey's Point retreat up river.

I'm talking to Tom Stephenson, still trying to figure the man out.

He gets asked a lot of questions these days. Questions like:

"Why don't you slow down a bit, Tom? You're not a young man anymore."

Or: "Don't you ever want to rest Tom? You don't have to work so hard."

Perhaps: "Tom, are you seriously considering doing another back flip off the back of that raft?"

He could have led a different life. He chose his. It's a lucky man, I suppose, who can say that.

Tom Stephenson in the Herb Shaw and Sons Limited's pole yard in Pembroke on June 5, 2008.
(Darren Calabrese / Courtesty of the Ottawa Sun*)*

But still, had he stayed a teacher at Algonquin College he would be retired today at maximum pension and benefits. He threw that aside to go back to the hurly, burly of the lumber camps. The dangerous life of a forest fire fighter. It's not a decision many people would make.

So I ask him, again, Why did you do it? And perhaps it is because this is our last day on the raft — or maybe just because we have spent enough time together — he pauses before answering. He's taking the question more seriously than most. Or at least more seriously than I've seen him take a question before.

"Funny thing was," he finally says, "I never could get enough of it."

"What do you mean?" I asked.

"Well, being outside with all the rivers and lakes, the forests, the fishing — I never wanted to leave this country. I saw people come here for their vacations, or go out on the river a few times a summer, and I never wanted to be them. Not even for one day.

"I never could get enough of it."

And when he says that, it occurs to me there are people you meet from time to time who genuinely love this country. Not because of something

The crew rows the square-timber crib past moored sailboats and into the Nepean Sailing Club as they complete their journey along the Ottawa River on Friday afternoon, June 27, 2008.
(Darren Calabrese / Courtesty of the Ottawa Sun)

they read. Or something they came to believe. But because they belong here.

Nepean Sailing Club, Ottawa, June 27, 2008

On our last day, Darren Calabrese took one of my favourite photographs of the trip. It was when we arrived at the Nepean Sailing Club where Dana had made arrangements to briefly moor the raft so it could be disassembled for the truck ride to the Canadian Museum of Civilization.

There was no way of getting directly to the museum. Dams and rapids blocked the way. So the sailing club would be our final stop. We drift along behind the sweep boat, taking a final look at the shoreline. The houses are numerous now. I spot familiar roadways.

Then, on a grassy knoll in front of us, we see people. We see them from a distance at first, and it is only when we get closer that we can see they are waving at us. Then it looks like a man waves to someone behind him, and the next thing we know, people are walking onto the knoll like it was some ridge a battalion company had just won.

"What is that?" I say to Dana.

"I'm guessing" he answers, "but I think that is the Nepean Sailing Club."

And sure enough it was, tucked behind the knoll where you couldn't see it from the main channel of the river. As we got closer to the sailing club, the crowd on the knoll grew, until Dave Lemkay yelled back at us that this is as close to the shore as he should go. We man the sweeps and start rowing the rest of the way.

When we get within fifty feet of the knoll, Paulette Gauthier, who has been on the sweep boat, grabs her fiddle and starts playing. A cheer goes up. We row past the knoll, then steer carefully into the small man-made basin that holds the sailing club, the strains of the "Log Driver's Waltz" ringing in the air.

That's when it happened: my favourite photograph of the trip. Right then, as we steer our way into the basin. There are a lot of sailboats at the Nepean Sailing Club, many of them of the near-yacht variety. Boats that are worth well into the six figures are moored in many of the slips. It is a busy spot, and space is at a premium.

So there we are, trying to steer the raft, hoping we don't crash into something, when we finally execute a near flawless left turn and start making our way past a line of boats. Behind us, standing on the knoll where he had asked to be dropped off, stood Darren Calabrese taking a photo.

When we saw it later that day, it looked like the boats were lined up to be inspected by this ragtag raft with the tree-limb flagpoles and the crew of lumbermen who hadn't shaved in a week. We sailed right between them, about as incongruous a vessel as the Nepean Sailing Club has probably ever seen. At the back of the raft you could see Tom Stephenson standing there like a captain, manning a rudder.

The raft was taken apart and put on the truck. It would be taken to the Canadian Museum of Civilization early Monday morning. A photo-op for local media had been scheduled for 10 a.m.

We shook hands and dispersed for the weekend, promising to be at the Waterfall Court no later than seven. Only when we were leaving did it occur to someone that we had never put the raft back together on dry land.

"You're right," said Dana. "I wonder if that's going to be a problem."

Good Question. See you Monday.

CHAPTER SIXTEEN

June 28–30, 2008

When Dana arrived home, he found newspapers waiting for him. The *Renfrew Mercury*. The *Eganville Leader*. The *Pembroke Observer*. The *Ottawa Sun*. He sat down at his kitchen table and read them, chuckling at the photo of Tom's friend, Scott McLellan, who had joined us for the last day on the raft, all six-foot-eight-and-more-than-three-hundred-pounds of him. Man that guy was big.

Some of the stories had mistakes (in one story he and Johnny had become brothers) but, for the most part, they got it right. The weight of the raft. The work that went into building it. The trip to Ottawa. Even the quotes from him and Tom. It was all there.

After the first day on the river, he got used to seeing reporters when they moored for the night. Even the CBC had come calling—inviting him and Tom to their studios in downtown Ottawa where they gave a lengthy radio interview. As for A-Channel television, it had practically been part of the crew, following them all the way to the Nepean Sailing Club.

Dana was pleased people had taken notice. The scrapbook his wife was putting together would be something the children might pass down to their children. Any man would be happy about that. But that wasn't the only reason.

He remembered, at the Nepean Sailing Club, an elderly woman came up to him and asked where they had stored the raft for the past 100 years. She didn't believe it had just been built.

"Where in the world did you find trees that big?" she asked.

Alice Township was the answer. Right outside Pembroke. And he went on to explain how loggers in Canada never depleted the forests the way they did in the States, where in a place like Pennsylvania you would be lucky to find white pine anywhere today—all hardwood, even though the forests there used to be nothing but pine and hemlock.

The Ottawa Valley was declared a conservation forest area 125 years ago when loggers in the States were almost clear-cutting their way across the northeastern seaboard. Even in Algonquin Park, which has always allowed logging, he would be surprised if any visitor ever noticed anything more than the odd logging truck. Or came away thinking the forests were in jeopardy.

When he was just a teenager, Dana had helped plant acres of red pine near the Shaw Woods, an area near the original mill on Lake Doré, where the

People congregate on a timber raft put together by the Gillies Lumber Company.
(Photo courtesy of the Arnprior and McNab/Braeside Archives)

family created a forest preserve in the seventies. There was original growth pine in the Shaw Woods. A butternut tree that was probably more than 200 years old. He went there sometimes to repair the walking path. Or inspect the trees.

As for the red pine he planted next to it, they had started harvesting that wood last year. God willing, he would see another cycle. Maybe even a third.

He took the newspapers and stacked them neatly in a pile, wondering for a moment how many people might come out to see them at the museum. There had been a lot of interest upriver where many people came from a logging background. Here in the Nation's Capital, at a museum that displayed

Egyptian treasures and Glenn Gould's piano, he wasn't sure.

Still, if even a few people came, he'd be happy. He had rented a hotel room across the street and would stay there with Tom for the week. As he made his way to bed, he wondered for a moment if Tom snored. Probably. That would be like him.

Tom Stephenson also came home to find a stack of newspapers waiting for him. He read them in his backyard where the tomato plants, after a brutal spring, were finally starting to grow.

The photos were almost comical. Scott looking as large as Big Joe Muffraw. Dana with his suspenders and Porky Pig hat. The tarp strung up and looking like the poorest seat on some run-down patio. He wondered if he could get an original somewhere. He might phone the *Pembroke Observer* when he got back from Ottawa.

He put aside the newspapers and walked around the backyard. The fruit trees had large flower buds, and he was pleased to see that. The raspberries and strawberries were doing well too. It would be late this year when the berries were finally ready to be picked, but they were coming along. Judy would make jams and preserves when they were ready—a few weeks later than normal—but when you spread it on a piece of toast in February, he doubted anyone would remember.

Most times, you can catch up. He learned that at the University of New Brunswick, arriving on campus with not even a high school diploma, so he started about as far back as a person could start. For the first couple years he barely slept at night, so worried was he about failure, about coming home and having to tell his family and the bank manager who was paying part of the tuition, that he couldn't do it. He had probably never worked so hard in his life, making sure that didn't happen.

Hard work gets you through most jams. He wondered why more people hadn't tweaked to that. The people who gave up when confronted with a tough road with a few obstacles. He never understood those people. Sure, you might fail. In this world anything was possible. But to not try. To curl up in a fetal ball. Might as well try. Only way you ever change anything.

He finished surveying the backyard, then took the stack of newspapers into the house. He was meeting Dana at the Museum of Civilization first thing

Monday morning, and there was plenty to do around the house before he left. Rafts should have gone downriver in August when there was less to do.

He put the newspapers away and then went downstairs. He should bring a few things with him to Ottawa. Now, where was the rope-making machine? The one he built after seeing a similar contraption at Upper Canada Village. It was here somewhere.

I went back home as well. Spent the weekend putting a pond in the backyard, momentarily giddy from the summer weather. Sunday evening, I was back at my desk, going through final notes, finishing the chap book about Philemon Wright.

Once again, I was struck by how courageous—and you have trouble writing that word today in this era of cynicism and diminished expectations—this man was.

Why are there no statues to Wright anywhere in the Nation's Capital? Simon de Bolivar: We have one of him. Ditto for forgotten cabinet ministers, female suffragettes, someone who drowned in the Ottawa River—but the man who first tried to live here, who made everything that came after him possible—nada. Not even a street is named for him.

After the raft trip of 1806, Wright built rafts and sent them downriver every year. Within a decade, others had arrived along the banks of the Ottawa River with their own land grants, ready to do the same. Within twenty years, the bend in the Ottawa River where three rivers meet had become something akin to the destination of a gold rush, a wild frontier place where brothels, taverns and desperate men outnumbered every legitimate business and honest citizen by a ratio of ten to one.

It must have broken his heart. This utopian farming community he envisioned, turned into a northern version of Deadwood.

As the years passed, he retreated more and more from the logging business. He helped establish St. James Anglican Church in Gatineau, the oldest church in the region, because he felt some sort of faith was needed in this New World, and although he had been a Congregationalist in New England, the Church of England would suit him fine here.

He petitioned the Royal Institution for the Advancement of Learning for a school. He founded a local agricultural society. Gave land (with the promise of proper compensation at some later date) to anyone willing to farm.

For the most part, however, he puttered in his gardens in his final years. Grafting potatoes in search of the perfect hybrid for this northern climate. Weeding his flower beds. Taking samples of wheat and sending them off for analysis in London.

He died on June 3, 1839, less than two months after that year's rafts had sailed for Québec. Within five years, both Tiberius and Christopher Columbus had also died.

Amazing. What happens to a man's plans when they are finally set forth in this world to compete in the ebb and flow of history.

I finish reading the book, set it aside, and look at the newspapers that have been piled on my desk. I read the stories that have been written about the raft trip. Laugh at the photo of the crib making its way past a line of boats at the Nepean Sailing club. I may frame that one.

Canadian Museum of Civilization, June 30, 2008

As I hurrry toward the Canadian Museum of Civilization — running late as usual, I'm about to get teased — I find myself remembering a time when no one thought this would work. Just a season ago. That's all it was.

Seems like a lifetime ago. Still a bit surprised we pulled this off. In modern-day Canada, no less. That strange place where we've spent three months waiting for Transport Canada to phone and tell us you can't sail rafts anymore. Or you need a bailing can.

But it never happened. Don't know if we'd be that lucky again. And then, to top it off, one of the finest museums in the country has welcomed us with open arms. Sent out press releases, no less. Even as I near the front door of the museum there is a part of me still waiting for some stranger to jump from a shadow somewhere and in the shrillest, bean-counting voice you can imagine, yell "Liability!" — and poof, like that, everything will fall apart.

After you send out a press release, it's too late. That's how it works, right?

I kept walking, thinking about the many obstacles the Shaws had to overcome to make this trip. The trip itself. And lastly, whether there are people who will actually come to see the raft, now that it is in Ottawa.

I had listened to Dana while he wondered the same thing at the Nepean Sailing Club, and found myself nodding. A lot of people upriver had been interested in the raft, but in a large city like Ottawa, where even in the Canadian Museum of Civilization this week you can see Glen Gould's piano and original artifacts from Québec, Jamestown and Santa Fe, well, it was difficult to say. We've never had that sort of competition before.

The raft, disassembled and sitting on a truck, arrives at the Canadian Museum of Civilization on the morning of June 30th, 2008. (Dana Shaw)

Just then I strode to the top of the stairs leading to the Waterfall Court. Saw the raft below me. I noticed right away how large it looked, now that it was out of the water. Even larger than in the pole yard, which had a lot of open space to surround it.

Then I noticed the Parliament Buildings in the background, the spires and copper roof reflecting the early morning sun. In front of Parliament Hill was the Ottawa River, with Ottawa Lockstation off to the left, boats tied up and waiting to go up the Rideau Canal. In the foreground of all that was the raft. It was a spectacular view.

But something else caught my attention. An image I'd like to leave you with. For already the raft was starting to attract attention. Below me I could see a security guard laughing at something Tom Stephenson was saying. Elderly couples with name tags on were milling around, looking anxiously at Dana as though waiting for a sign to approach. Cyclists stopped and stared. A young girl ran up to Johnny Shaw and pointed at a rope machine Tom had brought.

The raft being assembled at the Canadian Museum of Civilization.
(Dana Shaw)

Even as I stood there, a crowd started to form. By tomorrow—Canada Day—there would be tens of thousands of people milling around, each one, it seemed, delighted to discover a raft in their midst. Tom would invite people aboard. Dana Shaw would cooking meat for an impromptu barbecue. Scott McLellan would pose for photos.

And for that day, on this raft, you would have been hard pressed to find a single argument or discover people who did not get along. "What you want on a raft above all things," I remember Mark Twain writing, "is for everybody to be satisfied and feel right and kind toward the others."

Of course that's right. It's the kind of country we've been trying to build for two hundred years.

Wrightville, June 3, 1839

Philemon was hunched over in his gardens, weeding a row of potato plants. In the early years of the settlement he had set records in British North America for his potato crops. And his wheat. In 1823—he had read in his notes last night—the

Museumgoers and passersby enjoy the crib up-close after the square-timber reached its final destination in front of the Museum of Civilization Monday morning, June 30, 2008. (Darren Calabrese / Courtesty of the Ottawa Sun)

settlement had produced 71,630 bushels of wheat. And they had more fields of potatoes that year than they did wheat.

Still, there was too much wastage. On a bad year, an early frost would destroy the potato crop. He was not even convinced there had to be a frost. It seemed several cold nights in a row had the same effect.

To make the harvest predictable, stable, the plants needed to be heartier. He had already started experimenting with different kinds of potatoes, had requested seeds from the Royal Horticultural Society in London.

Was doing grafting of his own in the potting shed he had on his property.

He had read as many books about farming in British North America as he could find. Was particularly intrigued by the diaries of Champlain who was the first European to try and farm this New World. The great adventurer had encountered difficulties himself (most of his settlers ended up moving to Isle d'Orléans because the island in the middle of the St Lawrence was more temperate than the Habitation on top of the cliffs), but he persevered.

One plant Champlain had great success with was roses. He planted them throughout the Habitation. They still grew there, 200 years after his death. Philemon had been so fascinated by the story, he tried to grow the flower himself and found Chaplain was right. Certain species seemed to thrive in this climate.

He stood in front of some roses now. A giant bush of them growing next to the stone wall of his home. Different species bloomed at different times of the year, and if Philemon could figure it out properly, he could have roses blooming all spring, summer and fall. Could you imagine a rose growing in the snow? He had dreams about it. For several years now.

He didn't think it possible, but still, puttering in his gardens some mornings, the dream was so fresh he had trouble shaking it off.

He didn't spend much time on the timber business anymore. Certainly didn't travel to Québec (he lived in that city for months that first year, before finally selling his wood in early November). Those were jobs he left for his sons. They were prospering. As was anyone in the logging business these days. The rafts on the river every spring, there were so many of them it was a wonder there was any room left for the fish. The community across the river — Bytown it was called after the British Colonel who came in 1826 to build a canal — was thriving. More people arriving every day.

Philemon stood in front of the rose bush, and taking a pair of pruning scissors from a pocket of his pants, he gently cut off a full bloom. It was a vivid orange. The colour of a harvest moon. He cut four more blooms, enough to make a nice bouquet for his dining table, then put the scissors back in his pocket.

He made his way to a side entrance of his home. He had been feeling tired recently and thought he would have an afternoon nap.

J.M. Munro Sr. of the Munro Lumber Company (with pole) inspecting square-timber cribs at Seven-League Lake,
eighteen miles north of Mattawa, April 8, 1900

Epilogue

The sun has started to set over the Ottawa River, and the librarians left an hour ago. Only security guards are in the room now, and they came by, recently, with new rules (no digital cameras past 4:30 p.m.).

I have moved to a cubicle by the window. Below me the bike paths are filled with early evening cyclists, and I can see fishermen standing on the shoals off Victoria Island. It promises to be a beautiful summer night.

I move the box in front of me to one side. Inside, when I finally removed the lid earlier in the day, I had found twelve pocket-sized chap books. They were owned by Philemon Wright, with his signature on the inside cover of each one.

He called them his "raft" books. He started a new one each year in mid winter when he began chopping down trees for the rafts he would build in the spring.

Inside the books were his tallies of the trees felled, the names of the men who would make the raft trip that year and their wages. The dates of the voyages. How much pork and wheat each raft would need.

The cover of each book varied. You could see them getting more ornate as the years went by (Wright was becoming a prosperous man), and the titles varied as well. Some years the words "Raft Book" were prominent. Other years, the words "Orderly Day Book" were more prominent.

One thing that never changed was a date on the cover of each book. With one exception. That chap book was the smallest and poorest looking of the collection. The cover was frayed and tattered. There was no writing at all on the cover. You had to look inside to see the date—1806. The book Wright used for his first trip.

I imagined again what it must have been like. Felling trees in the dead of winter. Building nineteen rafts at the foot of the Chaudière Falls. Setting off down river, knowing he would have to pass the Chute de Blondeau, the Lachine Rapids, the Long Sault Rapids. How in the world did he ever think it would be possible?

Desperate men do desperate things. Perhaps there is no greater truth in the world. And although it was a trip fraught with danger and unexpected adversity, his timing turned out to be perfect. He had Napolean's blockade to thank for that.

A just reward? Cosmic coincidence? It is hard to say exactly what happened. Or maybe there was no connection at all, just history chugging along in its normal, sloppy way.

By now several weeks have passed since our raft trip. I talked to the Shaws and Tom Stephenson nearly every day afterwards as I began writing the summer series for the *Ottawa Sun*. Now though, I realized with surprise, it had been more than a week since I had spoken to any of them. Lives were being resumed. Days were falling back into a routine.

The raft itself went back on the water one last time in August as part of the annual summer festival in Pembroke. It travelled from Petawawa Point to Pembroke, and if we started out with seven people on the raft when we left the mouth of the Bonnechere River in June, there must have been closer to seventy for its final voyage.

John Yakabuski the provincial member of Parliament was aboard. Ditto for Cheryl Gallant the federal member of Parliament. Standing next to them were reporters, municipal officials and almost too many mayors to mention. From a project everyone laughed at not so long ago, it ended with an event you almost needed an invitation to get into.

I remembered Tom Stephenson's comment to me our last day on the river—"Never could get enough of it"—his explanation for why a man would want to spend his life in the bush, why he makes the decisions he does in this world. And Stephenson was right. Sometimes it's not about reason or planning. It's about something else.

"Never could get enough of it"—and I thought of Johnny Shaw reading his logging books, or Dana Shaw standing in the Shaw Woods pointing out the various trees. It was hard to imagine them doing anything else with their lives, or ever wanting to. On their last day on the planet, they likely would want a bit more time in the forest.

"Never could get enough of it"—and I remembered how panicked I was when I was running around trying to find someone to build a raft, dread-ing the phone call to an editor to tell him the story had just crashed and burned; then how excited I was when I finally stepped foot on the raft on the Bonnechere River. When we reached the Nepean Sailing Club, I didn't want to leave.

"Never could get enough of it"—and I thought of Philemon Wright tap-tapping his way down the frozen Ottawa River on his way to a New World with seven children in tow. When he had the opportunity to turn around and leave six years later, he set off on a raft trip, instead, that should have killed him.

It's hard to explain why people do the things they do, even in the best of times. But wanting more, it often has something to do with that. Maybe that's even what made Canada a great country. To travel down the river and see what's at the end, to build a city where no one thought it possible, to take a bunch of wonky rafts to Québec City. It was all about wanting more.

I take another look outside the window. The sun has set now and lights have been turned on along the bike path. I see cyclists flit in and out of the light, and for a moment, it reminds me of the timbers flitting in and out of the headlights on the Bonnechere River when we put the raft together that first day. Or the headlights on Sand Point.

In the end, I don't think it was love of history (although that was part of it) that made people fall in love with the raft. Nor was it just an off-beat story that momentarily caught their attention. I think, in that raft, people saw the "something more" they often look for in life, that thing that will take them from this place to the next.

A symbol? Sure, you could say it was a symbol. Of this country. The heyday of logging in the Ottawa Valley. Maybe even the Ottawa River. You can take your pick. But to be honest, I think most people saw it for exactly what it was—a thirty-ton square-timber raft. No symbols needed.

But when they saw it, something clicked inside. A memory was stirred. An emotion rediscovered. You looked at the thing, and you started to dream, started to think about what else in life was possible, what journey you could take next. I saw too many people walk away with smiles on their faces after seeing that raft to believe there was nothing personal about this. In that raft people saw a bit of themselves.

Saw the best parts. The crazy, dreaming, let's-give-this-a-try parts, the parts that have always produced something memorable and grand in this world. Like Philemon Wright's little settlement, or the country that sprang up around it.

I sort through the chap books on the desk, arranging them one more time in chronological order. There is talk of putting the raft on permanent display in Pembroke. Maybe down at the marina. It is a good idea, and I hope it happens, although I also hope it goes out on the river from time to time. A raft that doesn't go in the water? I've heard that debate before, and if you've read this book, you know how it was settled.

Wherever it is displayed, I know I'll be down to see it next year. Quite frankly, I never could get enough of it.

I pick up the chap book on the far left corner of the desk and open the cover. Start to read:

"11 July, took off with a raft bound and intended for Québec. Hands on board were Philemon Wright, Tiberius Wright . . . ."

I make myself comfortable. I am in no hurry.

Square timber rafts after they have completed their journey sitting in the Harbour of Québec City.

About the Author

Ron Corbett is an author, journalist and broadcaster living in Ottawa. Host of CFRA's popular evening show "Unscripted," he is also a columnist with the *Ottawa Sun*. Ron's previous book *The Last Guide* was a Canadian bestseller and he has won numerous awards for his writing, including two National Newspaper Awards.

He lives not far from the Rideau River with well-known photojournalist Julie Oliver and their children. He admits he always wanted to take a raft trip down the Ottawa River—just never thought he would do it on a square-timber crib.

ONE LAST RIVER RUN

Ron Corbett

TO ORDER MORE COPIES, CONTACT:

General Store Publishing House
499 O'Brien Road, Box 415
Renfrew, Ontario, Canada K7V 4A6
Tel 1-800-465-6072 • Fax 1-613-432-7184

www.gsph.com